Overcoming Difficulties with Number

Supporting Dyscalculia and Students who Struggle with Maths

Ronit Bird

Los Angeles | London | New Delhi
Singapore | Washington DC

SAGE Publications Ltd
1 Oliver's Yard
55 City Road
London EC1Y 1SP

SAGE Publications Inc
2455 Teller Road
Thousand Oaks, California 91320

SAGE Publications India Pvt Ltd
B 1/I 1 Mohan Cooperative Industrial Area
Mathura Road,
New Delhi 110 044

SAGE Publications Asia-Pacific Pte Ltd
3 Church Street
#10-04 Samsung Hub
Singapore 049483

Library of Congress Control Number: 2008941739

British Library Cataloguing in Publication data

A catalogue record for this book is available from the British Library

ISBN 978-1-84860-710-1
ISBN 978-1-84860-711-8(pbk)

Typeset by C&M Digitals (P) Ltd, Chennai, India
Printed in Great Britain by Ashford Colour Press Ltd.

MIX
Paper from
responsible sources
FSC FSC® C011748
www.fsc.org

Overcoming Difficulties with Number

Supporting Dyscalculia and Students who Struggle with Maths

Praise for this book:

'Ronit Bird is one of the most skilled and experienced teachers of learners suffering from dyscalculia. Her approach is based on years of reflective practice but also a deep understanding of the roots of numerical difficulties and disabilities. She stresses the importance of starting with concrete and manipulable materials before moving on to more symbolic materials. Her teaching scheme builds systematically on the basis of the learner's current understanding, rather than on mechanical measures of performance. This seems to me of fundamental importance. *Overcoming Difficulties with Number* provides a wealth of numerical activities and games, taking the most effective from a range of sources, including Cuisenaire rods and domino patterns for the earliest stages where learners are still counting in ones. As learners progress, clear methods for reasoning about more complex numbers are introduced. She provides very lucid methods for areas where many children, not just dyscalculics, have great difficulty, such as solving 5½ x 1½ or (x + 1)(x + 3) using grids. I highly recommend this book for teachers and teaching assistants who deal with children who have number troubles, but I also believe that most teachers of early maths will find much that is helpful with all learners.'

Professor Brian Butterworth, University College London

Contents

 Contents of CD

About the author

Ronit Bird is a teacher whose interest in pupils with specific learning difficulties began with a focus on dyslexia. She qualified as a teacher at London University and subsequently gained a further qualification as a specialist teacher. While working with dyslexic pupils in a mainstream school, Ronit started to develop strategies and teaching activities to help support the learning of pupils who were experiencing difficulties in maths.

Ronit has taught in both primary and secondary settings, and has worked as a special educational needs coordinator (SENCO) in both the independent and state sectors. She currently works as a teacher and as a contributor to professional development courses. Ronit also runs training courses on dyscalculia for Harrow subject leaders, teachers and teaching assistants as part of the Harrow Dyscalculia Project, and works in an advisory capacity with the participating schools.

How to use this book

This book is for anyone working with learners who have not yet gained a secure understanding of the key ideas behind addition, subtraction, multiplication and division. It is particularly targeted at older learners who could feel humiliated or embarrassed by having to work with material obviously intended for young children but who still need help with some of the fundamental numeracy concepts.

The book identifies a small core of key strategies for numeracy and provides a detailed and practical guide to teaching them, taking into account the prerequisite skills that underpin each strategy. The key strategies addressed are:

▶ Bridging through 10, and through multiples of 10.

▶ Subtraction as complementary addition, so that pupils are always working up (forwards).

▶ Empty number lines for both addition and subtraction, not only as a way of working but also as a way of supporting visualisation strategies.

▶ Visualisation techniques to support mental calculation.

▶ The area model of multiplication and division.

▶ Multiplication and division taught side by side, with pupils always working up (forwards).

▶ Using logic and reasoning to extend knowledge and proficiency.

In developing the approach featured in this book, I have deconstructed the essential teaching points of the numeracy strategies listed above and teased out a structured and logical sequence for teaching and learning. The resulting series of teaching activities provides for a systematic and cumulative progression in very small incremental steps, frequently reinforced. In this way, pupils can be introduced to a single new idea at every step, without being prematurely exposed to problems beyond their level of understanding. Pupils are therefore able to experience success while developing their mathematical understanding.

I make no attempt to devise a formal or prescriptive teaching programme. Instead, I have tried to record, in as detailed and accessible a manner as I can manage, ideas that I have found to be successful with my own pupils in combating a range of common misconceptions and difficulties. The suggestions that I offer are presented as a structured series of consecutive activities and the work is predominantly practical and oral. I prefer to avoid worksheets because I suspect that having pupils work through pages of written examples will only result in the pupils reinforcing the same inefficient strategies and bad habits that have contributed to their lack of progress. I strongly believe that the teacher is each pupil's most important resource and that the teaching focus should always be on developing logical thinking and mathematically sound cognitive models.

The activities and ideas in this book are all ready to use with a minimum of preparation. The only equipment needed is what is commonly available in the classroom or easily purchased: dominoes, dice, playing cards and number cards, Cuisenaire rods and base-ten blocks, paper and pencil. Various other resources, including printable game boards and information about Cuisenaire rods, can be found on the accompanying CD 💿.

The contents of the book are organised into four parts:

▶ Part I – How to help pupils stop counting in ones.

▶ Part II – The bridging technique.

▶ Part III – The area model of multiplication and division.

▶ Part IV – Reasoning strategies.

Each chapter begins with an overview of the subject, putting the teaching points into context. Each chapter also displays a summary of the individual steps that are later expanded into consecutive teaching activities.

Part I addresses one of the most common obstacles to struggling pupils' progress in numeracy, namely, their tendency to rely on counting in ones. A wealth of different ideas for activities and games are designed to promote component work and to help learners climb out of the 'counting trap'.

Part II contains a detailed step-by-step guide to teaching the bridging technique for both addition and subtraction. Following an analysis of the requisite pre-skills together with suggested activities for teaching these pre-skills to pupils for whom the concepts are not yet secure, two further chapters are devoted to the teaching of bridging through 10, and then through multiples of 10. These chapters advocate the teaching of subtraction as complementary addition, and explore how pupils can move from the concrete stage to the purely abstract stages of mental calculation.

Part III contains a detailed step-by-step guide to teaching the area model for multiplication and division. The first two chapters address both operations together and include an analysis of the requisite pre-skills complete with suggested activities for teaching these pre-skills to pupils for whom the concepts are not yet secure. Two further chapters are devoted to the teaching steps though which pupils can learn to manage the transition from the concrete stage to the abstract stage of understanding the standard written algorithms.

Part IV focuses on reasoning strategies. Pupils with difficulties in maths are rarely flexible in their thinking and must be explicitly taught how to use the few facts they know to derive new facts.

The CD accompanying this book contains a short introduction to Cuisenaire rods as well as printable resources for games, puzzles and activities 💿.

Background

Pupils underachieve in maths for all sorts of reasons. One of those reasons might be teaching methods that are not well matched to the pupils' learning needs; another might be the existence of learning disabilities. The Williams Review into maths teaching and learning in UK primary schools (DCSF 2008) gives a figure of approximately 6 per cent of pupils who have fallen significantly behind (i.e. who fail to achieve level 3 of the National Curriculum) by the end of Key Stage 2, the stage at which children transfer from primary to secondary school, and points out that this percentage has remained roughly constant for almost a decade.[1] Thus, despite the introduction of the National Numeracy Strategy in the late 1990s and the numerous government initiatives directed at improving maths teaching and learning in schools over the past decade, there is still a small number of learners who have significant and persistent difficulties with learning basic numeracy.

At the same time, there is growing evidence of specific learning difficulties affecting maths. Professor Brian Butterworth, Britain's leading authority on dyscalculia, estimates that 4–6 per cent of the population have dyscalculia, a disorder that is neurological in origin and that results in difficulties in learning about number and arithmetic.[2] Those working in the field of dyslexia estimate that 4–10 per cent of the general population are dyslexic;[3] of these roughly 50–60 per cent are thought to have associated difficulties with aspects of mathematics.[4] These types of specific learning difficulties are found in learners of any cognitive ability.

These different sets of figures are consistent with one another. They suggest that teachers should expect to find one or two pupils in every average class for whom maths will be a struggle. Whatever the cause of their difficulty, pupils who have not mastered basic numeracy concepts will find it impossible to make satisfactory progress at secondary school without specialised teaching to address their particular problems.

Specific learning difficulties affecting numeracy

Dyscalculia

Developmental dyscalculia was first recognised by the Department for Education and Skills (DfES) in 2001[5] and defined as:

> a condition that affects the ability to acquire arithmetical skills. Dyscalculic learners may have difficulty understanding simple number concepts, lack an intuitive grasp of numbers,

and have problems learning number facts and procedures. Even if they produce a correct answer or use a correct method, they may do so mechanically and without confidence.

As a teacher, you might suspect that you have a dyscalculic pupil in your class if an otherwise competent student has a surprising level of difficulty with ordinary numeric operations and relies on finger-counting, often for all four arithmetic operations, well beyond the age at which most of the others in the class have progressed to more efficient strategies. A dyscalculic learner stands out as having no 'feel for numbers' at all, no ability to estimate even small quantities, and no idea whether an answer to an arithmetic problem is reasonable or not. Memory weaknesses, both long-term and short-term, are a great handicap and result in a pupil with dyscalculia being unable to remember facts and procedures accurately, or consistently, no matter how many times they try to learn them by heart. Pupils who have dyscalculia simply cannot remember their times tables reliably, and you may find they can recall some facts one day but not the next. They are also likely to lose track of what they are doing when attempting any procedure that requires more than two or three steps. Even basic counting can be a problem for pupils with dyscalculia, especially counting backwards.

Indicators for dyscalculia are:

▶ An inability to subitise (perceive without counting) even very small quantities.

▶ An inability to estimate whether a numerical answer is reasonable.

▶ Weaknesses in both short-term and long-term memory.

▶ An inability to count backwards reliably.

▶ A weakness in visual and spatial orientation.

▶ Directional (left/right) confusion.

▶ Slow processing speeds when engaged in maths activities.

▶ Trouble with sequencing.

▶ A tendency not to notice patterns.

▶ A problem with all aspects of money.

▶ A marked delay in learning to read a clock to tell the time.

▶ An inability to manage time in their daily lives.

Dyslexia

A dyslexic pupil might show many of the same indicators as those mentioned above, because it is thought that at least half of all dyslexics also have difficulties with maths. Outside the maths classroom, you might suspect that pupils are dyslexic if they read and write much less willingly and fluently than you might expect, if they read and reread written material with little comprehension and if their spelling is particularly weak, inconsistent or bizarre. Dyslexic learners show much greater ability and understanding when speaking than you could ever guess from looking at the scrappy and minimal amount of written work they produce. Other indicators are memory weaknesses, problems with processing auditory information, and difficulties with planning and organisation.

Dyspraxia

A typical dyspraxic pupil does not seem to have the same long-term memory problems as a dyslexic and so might be able to remember times tables facts with ease. Dyspraxia, also known as DCD (developmental coordination disorder), mainly affects motor control, which results in pupils being clumsy and uncoordinated, poor at planning and organisation, and unsuccessful at subjects like PE and sports that require balance and coordination. Dyspraxic pupils cannot process sensory information properly and are therefore forever tripping and falling, dropping and breaking things, and mislaying their belongings. In the maths classroom, dyspraxic pupils have particular difficulty handling equipment such as a ruler, a protractor or a set of compasses, and their written work is likely to be very messy and difficult to decipher.

Diagnosis

A quick and informal way of identifying pupils who need extra help, or further assessment, is to: (a) find whether pupils have difficulties counting backwards, (b) discover which pupils cannot remember times tables reliably, and (c) notice which pupils have no calculation strategies beyond counting in ones. A less subjective identification can be achieved by using Brian Butterworth's computer-based Dyscalculia Screener, published under the nferNelson imprint and obtainable through GL Assessment. The Screener, which is based on Professor Butterworth's neuroscientific research, can be administered to several pupils at once and produces a profile of each pupil that can provide evidence (or an absence of evidence) of dyscalculia. A formal diagnosis of dyscalculia can, like dyslexia or dyspraxia, only be given by a qualified educational psychologist after a thorough assessment.

How to help pupils who have difficulties with numeracy

The following principles summarise this book's approach to teaching and learning the key numeracy strategies.

▶ Start with concrete materials, making sure that the equipment you use is mathematically sound and is robust enough to model a wide range of numeracy topics. In my opinion, the best concrete resource is a collection of base-ten materials, such as Cuisenaire rods, supplemented by Dienes blocks.

▶ Allow the pupils to use the concrete materials themselves. Do not appropriate them solely for demonstration purposes.

▶ Never allow the concrete materials to be used mechanically, simply to find an answer. Their value lies in the way they can be used to support visualisation techniques and to build cognitive models.

▶ Target pupils who are using counting as their only calculation strategy. Before mathematical progress can be made, pupils must be helped out of the 'counting trap' by learning to think in terms of components, or chunks, for building or partitioning numbers.

▶ Allow plenty of time. This means allowing pupils as much time as they need to use concrete materials and to experiment with them. It means building into your teaching plenty of opportunities for recap and revision. It means pausing after asking a question so that a pupil has enough time to think about what the question means without feeling rushed, and yet more time to come up with a reasoned answer. It means addressing problems and misconceptions not at the end of the lesson or the end of a topic, but when there is still time enough for pupils to reconstruct their understanding. It also means removing the time pressure of having to get through a whole term's syllabus in a single term and instead allowing pupils to become thoroughly familiar and secure with one topic before moving on to the next.

▶ Engage in a lot of talk as you work. Encourage the pupils to talk about what they are doing at every stage. Get pupils into the habit of reflecting on what they do or see and on putting their thoughts into words. Use and teach a wide variety of appropriate vocabulary.

▶ Focus on practical activities and teaching games. Apart from the fun that is to be had from solving puzzles and playing games, let your pupils see that maths is something we do, not something we necessarily need to write down. Introduce written calculation as a way of recording only what the pupil has already done concretely and/or has already understood.

▶ Progress in very small steps. Break down the teaching and learning of every topic into tiny incremental steps and address only one new idea at a time.

▶ Aim to move pupils gradually from the concrete stage through the diagrammatic stage before moving to the purely abstract stage of calculation.

▶ Lighten the burden on pupils' working memory by encouraging them to minimise the number of steps in any calculation.

▶ Give more attention to how a solution is reached than to what the solution is.

▶ Let pupils make mistakes. Encourage them to see errors as not only inevitable but also a helpful part of the learning process.

▶ Teach each new strategy by building on a foundation of what is already known. Check that all the necessary pre-skills are secure. Make the connections explicit.

▶ Allow informal calculation strategies to replace standard written algorithms provided that pupils can consistently reach the correct solution in a reasonable amount of time.

▶ Minimise the amount that pupils are expected to commit to long-term memory by focusing on key strategies, i.e. those having the widest application. If there are several acceptable ways of tackling or recording a calculation, do not expect your pupils to become familiar with them all. Instead, allow individual pupils to choose whichever method they are most comfortable with and encourage them to practise that method consistently.

▶ Teach reasoning strategies explicitly. Show pupils how to use logic and reasoning to extend their knowledge of facts and procedures.

References

1   Department for Children, Schools and Families (DCSF) (2008) *Independent Review of Mathematics Teaching in Early Years Settings and Primary Schools*. Ref: DCSF–00433–2008.
2   B. Butterworth (2004) 'Developmental dyscalculia', in *The Handbook of Mathematical Cognition*, vol. 1, 6, ch. 26, ed. J.E.D. Campbell, p. 455, Routledge.
3   Department for Innovation, Universities and Skills (a government department set up in June 2007), www.dfes.gov.uk/curriculum_literacy/access/dyslexia/
4   L.S. Joffe (1981) 'School mathematics and dyslexia', unpublished doctoral thesis, University of Aston, Birmingham, UK.
5   Department for Education and Skills (DfES) (2001) *Guidance to Support Pupils with Dyslexia and Dyscalculia*. Ref: DfES–0512–2001.

PART I

How to help pupils stop counting in ones

CHAPTER ONE

More than 50 ideas to help pupils stop counting in ones

Overview

Many pupils who struggle with arithmetic have a tendency to count in ones. What is a normal stage of development for most children becomes a crutch for pupils with poor number sense. Pupils who continue to rely on this unsophisticated and laborious strategy well beyond the stage at which counting is appropriate or efficient have fallen into the 'counting trap'.

The counting trap is the situation in which pupils know very few arithmetic facts for certain, and therefore have to calculate every new fact from scratch. They calculate by counting in ones, an arduous and long-winded process that puts so much strain on their already weak memories that the newly found answer becomes dissociated from the question and therefore cannot be added to the store of known facts. Which, in turn, means that very few number facts can be instantly recalled or relied upon, so that every new fact must be calculated afresh.

In order to help such pupils make progress, it is essential to teach them to replace their ones-based approach with chunking techniques. The aim is to minimise the number of calculation steps in order to increase a pupil's chances of achieving a correct solution in a reasonable amount of time and without putting any undue strain on working memory. Eddie Gray, writing in *Teaching & Learning Early Number*,[1] explains how a reliance on counting inhibits flexibility and puts forward the idea that in order for newly calculated facts to be laid down in long-term memory the counting process must be compressed.

Pupils who habitually count on their fingers need to be given lots of opportunities to work through targeted activities that help them become thoroughly familiar with the number bonds of the first ten whole numbers, and then, by extension, of all the whole numbers up to twenty. Pupils must also engage in activities that encourage them to partition numbers into suitable components and to manipulate the components, rather than collections of single units, as they work towards a solution. The objective is for pupils to replace their immature habits with more efficient methods. However, pupils will, understandably, be reluctant to relinquish their well-established counting habits before they feel absolutely secure about any new approach. It is only after plenty of practice, therefore, that pupils will begin to accept that working with components is better than working with ones.

Games and puzzles are by far the most enjoyable way of getting the necessary practice in component work. This chapter contains more than fifty suggestions for suitable games and activities designed to appeal to adolescent learners. Pupils can be introduced to the games alongside their work on the bridging techniques described in Part II, and should, of course, accept that counting in ones is forbidden during play.

The ideas in this chapter encompass a variety of teaching games and activities targeted at various stages of learning. Included are games that exploit representations of discrete materials arranged into patterns, such as dominoes and dice, games that incorporate continuous concrete materials, such as Cuisenaire rods, and purely abstract activities and puzzles that require only cards or paper and pencil. Pupils should be encouraged to play games from each category. The abundance and variety of the ideas enable new games to be introduced frequently, together with new variations of familiar games, so as to provide regular revision without undue repetition.

The ideas can be used with individuals or with groups of pupils who are being withdrawn from class in order to help them understand the essential arithmetic groundwork without which they will not be able to access the rest of the maths curriculum.

The ideas are also ideal for extra-curricular maths clubs.

Summary of the component games in this chapter

Component game	Number of players	Equipment required
Whose Number Wins?	2	4 sets of domino cards* (20 cards each)
Snap	2	4 sets of domino cards* (20 cards each)
Pelmanism	2	2 sets of domino cards* (20 cards total)
All in a Row	1, 2 or 3	1, 2 or 3 sets of domino cards* (10 cards each)
3-in-a-row Key Components	2	Paper and pencil, one 6-sided die
Basic Domino Game	2 or more	Set of 28 dominoes
Make a 1–10 Sequence	1, 2 or 3	Set of 28 dominoes
Threes and Fives	2	Set of 28 dominoes
Round the Spot	Any number	Three 6-sided dice
Centenniel	Up to 5	Three 6-sided dice
Triples Addition	2 or 3	Three 6-sided dice, or 10- or 20-sided dice
Odds or Evens	2 or 3	Three 6-sided dice, or 10- or 20-sided dice
Stuck in the Mud	2 or 3	Five 6-sided dice
Three of a Kind	2 or 3	Five 6-sided dice
Who Has the Last Word?	2	Cuisenaire rods, one 10- or 20-sided die
The 3-Component Challenge	Any number	25 specified Cuisenaire rods each
Cherry Picking	2	Cuisenaire rods, one 10-sided die
Rods in Blocks	2 or 3	Cuisenaire rods, specially labelled die, paper
Descent	2 or 3	Cuisenaire rods, 6-sided die, paper tray
Tens Away	1	Pack of playing cards or digit cards*
Eleven Up	1	Pack of playing cards
Thirteens & Fifteens	1	Pack of playing cards

(Continued)

(Continued)

Fifteen in a Suit	1	Pack of playing cards
Standing Aces	1	Pack of playing cards
Pyramid Solitaire	1	Pack of playing cards
Prisoners	1	Pack of playing cards
Pontoon	3 or more	Pack of playing cards
Zero Blackjack	3 or more	Pack of playing cards
Shut the Box	2 or 3	Playing cards or digit cards,* two 6-sided dice
Banking Tens	2 or 3	One or two packs of digit cards*
Marching On/Marching Back	2	Playing board,* digit cards,* spinner
Conjure the Number	2	Pack of digit cards*
Subtract from 15	2	Pack of digit cards*
Minimise the Difference	2	Pack of digit cards,* paper and pencil
Maximise the Difference	2	Pack of digit cards,* paper and pencil
Plus or Minus	2	Plus or Minus cards* pre-prepared by pupils
Magic Squares and Number Puzzles	Any number	Paper and pencil
Component Su Doku	Any number	Su Doku puzzles at various levels,* pencil

* Can be printed off from the CD.

Dice and domino patterns

Why use them?

Amounts that are otherwise too large to subitise (that is, quantify at a glance) can be read without having to count in ones if the discrete items are arranged in a visually recognisable pattern. Dice and domino patterns are visual patterns that can be easily recognised and are well known by most people.

The dice patterns from 1 to 6 are shown here. It does not matter if the 2 and 3 are sometimes represented vertically or horizontally, rather than diagonally. Amounts as small as these can be readily subitised; numbers above 4 cannot usually be subitised, except when the units are arranged in recognisable patterns.

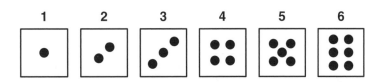

These patterns can be extended for numbers up to 10 by making doubles patterns for the even numbers and near-doubles patterns for the odd numbers. Doubles patterns are patterns that highlight two identical components, i.e. showing 8 as being built from two patterns of 4, rather than from, say, 6 and 2.

Some numeracy activities based on these dice patterns can be found in my book *The Dyscalculia Toolkit.*[2] Many other ideas for young children can be found in *Dyscalculia Guidance* by Brian Butterworth and Dorian Yeo.[3] A variety of commercially produced games, such as board games in which dice are used, or the dice game Yahtzee, can provide practice in subitising and dot pattern recognition. Several quick and simple dice games suitable for older pupils are described below.

Pupils of secondary school age who perceive dot patterns as childish often prefer to work with dominoes. Dominoes feature in several of the activities, puzzles and games below. As well as clearly showing the doubles and near-doubles components of the numbers up to twelve, dominoes are a fruitful resource because they also show other ways of splitting numbers into paired components.

Activity

Make domino pattern cards for the numbers 1–10

Pupils should start by building all the even numbers up to 10, by arranging counters or nuggets in doubles patterns on top of a domino background that has been created by splitting a rectangle into two squares or oblongs. Pupils can then make their own set of domino cards by sticking small round labels onto rectangles of card, or by drawing spot patterns onto paper or card.

Pupils should be able to read these domino patterns in two ways, both as a total and as a doubles fact. For example, they should recognise the last card shown here as '10' and also as '5 and 5'.

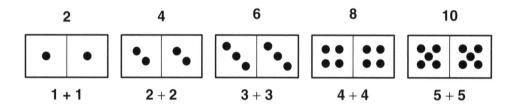

Next, have pupils build all the odd numbers up to 10 out of counters or nuggets, by replicating one of the two patterns from each of the adjacent doubles layout. For example, when making the number 7, position the new domino outline between the 6 and 8 patterns and build the new pattern out of half the 6 and half the 8, i.e. 3 + 4.

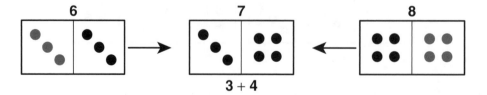

(Continued)

(Continued)

Pupils should make cards for all five of the even numbers first, and then all five of the odd numbers.

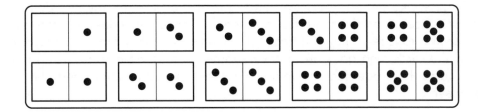

You will find domino pattern cards on the CD 🖸 ready to be printed off, but be aware that pupils can learn a great deal from making their own cards.

Play component games with domino cards

Two of the games below can be played with one set of 10 cards for each player. Other games require two sets of cards for each player.

Whose Number Wins?

Two players play with two sets of domino cards each, i.e. 20 cards each.

Rules: Players turn over one domino card at a time, simultaneously, from their own shuffled pack. Each player reads aloud the total number of spots on his/her own card. The player with the higher number wins both cards. If both cards are the same, they remain on the table to be appropriated by the winner of the next round.

Variation: Play so that the lower number wins.

Snap

Two players play with two sets of domino cards each, i.e. 20 cards each.

Rules: Players turn over one domino card at a time, simultaneously, from their own shuffled pack. If both cards show the same total number of spots, the first player who calls out 'snap' wins all the face-up cards.

Pelmanism

Two players play with two sets of domino cards, i.e. 20 cards in total, shuffled and spread out face down on the table.

Rules: Players take turns to turn over and look at the faces of any two domino cards, which remain face up on the table for the other player to see while the total quantity of spots on each card is read aloud. If both total numbers are the same, the player removes and keeps the cards, and goes on to have another turn. If the card totals do not match, the player turns them both face down again in their original positions and play passes to the next player.

All in a Row

Two or three players each play with a set of 10 domino cards. Alternatively, it can be played as a solitaire game in which players try to beat their previous record.

Rules: Each player shuffles their 10 domino cards and turns over the top five cards, arranging them in order of magnitude. The idea is to see how many cards show adjacent numbers. Score two points for a run of 4 adjacent numbers and ten points for a run of 5 adjacent numbers. The winner is the player with the highest score after 5 rounds.

3-in-a-row Key Components

Two players each sketch a set of domino patterns of the doubles and near-doubles facts, on squared paper, leaving the spots as empty circles, as shown below.

N.B. Do not be tempted to provide a ready-made template for this game, as the preparation makes pupils actively notice the doubles and near-doubles patterns.

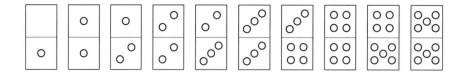

Rules: Players take turns to throw a die. A throw of 6 results in the player missing the turn. A throw of any other number allows the player to shade in the matching spots in any *one* place that the pattern appears. For example, if your dice throw is 4, you may choose to colour in the pattern of four on one half of the 8 domino, or on one section of the 7 or 9 dominoes. The winner is the first player to shade in all the spots on three consecutive dominoes, i.e. the first to complete 3 numbers in a row.

Play component games with a set of dominoes

A full set of European dominoes has 28 tiles or stones, from double-zero to double-six, with all the combinations in between.

Whose Number Wins, Snap, Pelmanism, All in a Row

These four games, described above as games to play with domino cards, can also be played with real dominoes. Whereas playing with the cards results in pupils focusing on the doubles and near-doubles component facts, playing with a full set of dominoes provides practice in the other component facts.

Basic Domino Game

This traditional basic domino game is for two or more players using a full set of 28 dominoes.

Rules: Players take the number of stones that results from the calculation '8 minus the number of players'. After an initial play of a double (6–6 takes precedence, then 5–5, etc.), take turns to place stones in a single straight line, matching the number of spots touching an adjacent stone. You must take a stone from the boneyard if you cannot play and must continue until you can play, provided that at least two stones remain in the boneyard until the end of the game. Play stops when either a player has used all his/her stones or no one can place a domino. The number of dominoes remaining in each player's hand is counted and awarded to the player who finished first. The winner has the highest score after 3 or 5 rounds.

Variations: Another way of scoring is for all the spots on each player's remaining dominoes to count against them, in which case the winner has the lowest score after 3 or 5 rounds. One common variation of the game is for any player placing a double to get an extra turn; another is to dispense with the boneyard and share all 28 stones between 4 players.

Make a 1–10 Sequence

This game for two or three players is played with a full set of 28 dominoes. Alternatively, it can be played as a solitaire game in which players try to beat their previous record.

Rules: Players turn over dominoes one at a time and keep a record of how many they turn over before uncovering enough stones to represent each of the numbers between 1 and 10 inclusive. The winner has the lowest score after 5 rounds.

Threes and Fives

In this traditional domino game, two players use a full set of 28 dominoes.

Rules: Players take 7 stones each. After an initial play of a double (6–6 takes precedence, then 5–5, etc.), take turns to place stones in a cross shape, with stones allowed from the middle of this first stone only, in both directions as well as from the ends. If you cannot play on your turn you must take one or, if necessary, two new stones from the boneyard, which you must play immediately if you can, or add to your stock if you cannot. The aim is to place a stone so that the free ends add up to a multiple of three or five, in order to score the amount of the multiple. Although stones may be placed from the middle of the 6–6 domino, only the patterns at the ends of the dominoes count towards a total.

For example, in the game shown below at the left, the player who places the 6–1 stone scores 4, because the free ends total 12, and the player about to place the 6–4 stone will score 2, because the free ends will total 10. In the game at the right, the total of the free ends will be 15 after placement of the 6–4 stone resulting in a score of 8 (since 15 is a multiple of both 5 and 3). As soon as one player has used all his/her stones the game is over, and the spots on the other player's remaining stones are added to count as penalty points.

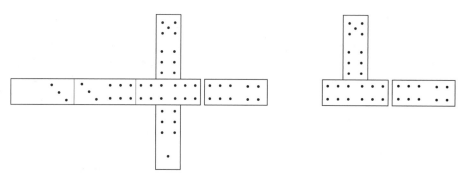

The five stones at the left have been played, scoring the last player 4 (i.e. 12 ÷ 3). The next player places the 6–4 stone at the right for a score of 2 (i.e. 10 ÷ 5).

Placing the 6–4 stone in this game produces a total of 15, scoring 8 (i.e. 15 ÷ 3 + 15 ÷ 5).

Play component games with dice

Five of the following six dice games are addition games. Players may not find their scores by counting up in ones, on their fingers or otherwise, since this is exactly what we are trying to eradicate by encouraging pupils to play these games.

Round the Spot

Any number of players can play this traditional dice game with three 6-sided dice.

Rules: Players take turns to throw all three dice at once. The name, 'Round the Spot', refers to the dice patterns that are clustered around a central spot, i.e. the odd numbers. A number 5 scores four points (four 'petals' around the central spot), a 3 scores two points; however, note that a 1 scores one point. After every turn, each player adds and records his/her own score. Even

(Continued)

(Continued)

numbers carry no score, but if a throw shows even numbers on all three dice, the player scores double what he/she scored on the previous turn. The game ends when all the players have had four turns each, and the winner is the player with the highest cumulative score.

Centenniel

This traditional dice game is for up to five players using three 6-sided dice.

Rules: Each player writes the numbers from 1 to 12 on a piece of paper. Players take turns to throw all three dice at a time. Players cross off the numbers from their list in numerical order. More than one number can be crossed off on a turn, but no number can be crossed off before the previous number has been eliminated. For example, a first throw of 2, 3 and 6 means the player cannot go, while a throw of 2, 3 and 1 allows all three numbers to be eliminated.

Variation: A common version of this game is for players to have to eliminate all the numbers in turn up to 12, and then all the numbers back from 12 to 1. In this variation, a throw of three 6s after the number 11 has been crossed off is treated as a special case that allows both 12s to be eliminated at once.

Triples Addition

This simple dice game is for two or three players using three 6-sided dice.

Rules: Each player takes turns to throw the dice as follows: throw all three, leave the die showing the largest amount and throw the remaining pair; leave the die showing the larger amount and throw the other. Find the total of the three dice and record the score. Play now passes to the next player. The winner is the player with the highest total after 5 rounds each.

Variation: For practice in adding larger numbers, use 10-sided or 20-sided dice, or a mixture of the two.

Odds or Evens

This simple addition game is for two or three players using four 6-sided dice.

Rules: Each player takes turns to throw all four dice at once, but before the throw must choose either odds or evens, and announce the choice to the other player(s). On your throw, you may add only the numbers that conform to your stated choice. The winner is the player with the largest total after 5 rounds.

Variation: For practice in adding larger numbers, use four 10-sided or 20-sided dice, or a mixture of the two.

Stuck in the Mud

This traditional addition game is for two or three players using five 6-sided dice.

Rules: Each player takes turns to throw the dice as follows: take any dice showing the numbers 1 or 2 out of play (they are 'stuck in the mud'), add the numbers on any remaining dice, then throw these remaining dice again and follow the same procedure until all five dice are 'stuck in the mud'. Keep a running total mentally and record the score as a single total at the end of each round. Play then passes to the next player. The winner is the player with the highest total after 5 rounds.

Three of a Kind

This addition/multiplication game is for two or three players using five 6-sided dice.

Rules: Each player takes turns to throw all five dice together. On your throw, if the dice show three identical numbers, you score the total for the three numbers. If four of the dice show the same number, you score not only the total of the four dice, but also an extra bonus of 5 points. If all five dice show the same number, you score an extra bonus of 10 points, in addition to the total of the five dice. The winner is the player with the highest total after 10 rounds of the game. Since many rounds will generate zero scores, you may prefer to end the game only when any one player has had 5 scoring rounds.

Cuisenaire rods

Why use them?

Although dice patterns and domino spot patterns are good for promoting visual recognition of component facts, the spot patterns are all built from discrete units. Over-reliance on such patterns will inevitably encourage pupils to continue thinking about numbers as composed of quantities of ones, which is precisely what we are trying to discourage. Cuisenaire rods, on the other hand, show each of the numbers 1 to 10 as a single unit. For example, the black rod representing the number seven can not only be measured in ones to demonstrate that it is equivalent to seven ones, but is also a discrete unit in its own right, i.e. one seven.

I find Cuisenaire rods to be an invaluable resource, responsible for many of those 'aha!' moments when suddenly things seem to click into place in the learner's mind. As with all other concrete teaching materials, rods should never be used to find the answer to a calculation mechanically, but, instead, be used as tools for modelling arithmetic thinking. Many ideas for exploring basic number concepts through rods can be found in *The Dyscalculia Toolkit*.[2] Many more can be found in numerous articles by Professor Sharma.[4] A short introduction to Cuisenaire rods, first printed in *The Dyscalculia Toolkit*, is reproduced on the CD.

Play component games with Cuisenaire rods

The six games below all require a set of Cuisenaire rods. A single mini-set, sold in a rigid black plastic box with a red lid, contains enough equipment for two players.

Who Has the Last Word?

This game is for two players. The game not only teaches component pairs, but also highlights the fact that there are a limited number of ways in which numbers can be split into components. This is especially striking with smaller numbers. For example, each of the numbers 2 and 3 can be split in only one way, i.e. 2 = 1 + 1 and 3 = 1 + 2.

Rules: Use a 1–10 die and throw again if the throw is 1. One player starts by throwing the die and taking a single rod to match the throw. Players now take turns to build the same number out of two components. If you are the first player, you must show the double or near-double components of the number by putting two rods end to end below the rod matching the die throw, while explaining the relationship aloud.

For example:

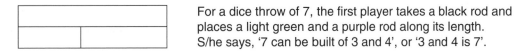

For a dice throw of 7, the first player takes a black rod and places a light green and a purple rod along its length. S/he says, '7 can be built of 3 and 4', or '3 and 4 is 7'.

The next player must make another line of two rods, showing a different pair of components (4 + 3 is the same as 3 + 4, so does not count as a new combination). Play continues until a player cannot make a new row, either because he/she doesn't know the number bonds very well, or because the possibilities have run out. The player who has had the 'last word' gets to keep the last pair of rods until the end of the game. After an agreed number of turns, the winner is the player who has won the largest total amount in rods.

Variation: Play with two ordinary 6-sided dice for numbers up to 12, or with two 10-sided dice for numbers up to 20. Players must still start with the double or near-double fact about the number, and not with the components shown on the two dice.

The 3-Component Challenge

This game, giving practice in combining three components, is played as a race against time between members of a group of any size. Players should each have access to their own allocation of 25 Cuisenaire rods comprising 4 each of the colours white, red, light green, purple and yellow and one each of the remaining five colours to represent the target numbers. No other rods are allowed. Players should also have a screen, e.g. a large book or a box file,

(Continued)

(Continued)

behind which they can experiment without other players being able to see and copy what they are doing.

The first challenge is to be the first player to succeed in building the five target numbers from 6 to 10 inclusive out of three, and only three, rods.

A more difficult challenge introduces the additional constraint that each of the target numbers must be built out of three rods of three different colours. Players soon discover that there is only one way to build the numbers 6 and 7 out of three different components, and will need to experiment with the numbers 8, 9 and 10 to find the unique solution to this second challenge.

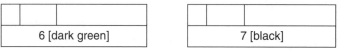

There is only one way to build the numbers 6 and 7 from three different components.

Cherry Picking

This game, for two players, gives practice in building single-digit numbers out of smaller components. As being the first to play gives an advantage in this game, players should take turns to start. Before play begins, the players arrange two of each colour of Cuisenaire rods, in order of size, flat in the lid of the rod box, with the lid placed between the players. The rods can be conveniently arranged either as a double staircase or as a wall, the same arrangement of twenty rods being shared by both players.

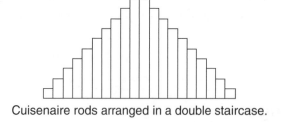

Cuisenaire rods arranged in a double staircase.   Cuisenaire rods in a wall.

Rules: Players take turns to throw a 1–10 die and take out of the lid as many rods as possible that add up to the exact value of the number thown. Each rod taken counts towards your score at the end of the game. If you cannot match the dice throw on your turn, you have to miss that turn. The game ends when both players, between them, have to miss three consecutive turns. Score one point for each rod you have removed, irrespective of its length or value. The winner is the player with the highest score after 4 rounds.

Rods in Blocks

This game for two or three players provides practice in splitting numbers greater than 5 into smaller components. Each player prepares for the game by drawing a rectangle measuring 5 cm by 10 cm on 1 centimetre squared paper. Leave a small margin of between half and one centimetre around the edges of the rectangle so that when it is cut out it is possible to fold up and secure the extra margins at the edges to produce a shallow paper tray of 5 × 10 cm.

A more robust alternative is to cut a bespoke shape from craft foam, or packaging foam, leaving a frame of a few centimetres surrounding a rectangular hole measuring 5 by 10 centimetres.

Rules: Players take turns to throw a 6-sided die labelled with the numbers 6 to 11 inclusive. On your turn, you must take rods to match the throw in the most efficient way possible, i.e. taking as few rods as possible, with the aim of eventually blocking out the whole of your rectangle with rods. Place your rods along the side measuring 5 squares and do not start a new column until the previous column is full. For example, an initial throw of 6 must be split into 5 + 1 (a yellow and a white rod) and a second throw of 11 must be split into 4 + 5 + 2 and arranged as shown below. The winner is the first to block out their entire space with rods.

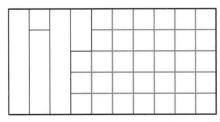

The position in the game Rods in Blocks after two dice throws: first 6 then 11.

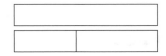

A third throw of 7 would need to be split into 3 and 4.

Variation: Use a die labelled 11–16 on a square measuring 10 cm by 10 cm.

Descent

This subtraction game for two or three players provides practice in decomposing tens into units, as well as in partitioning single-digit numbers into component parts. Each player starts by making a shallow paper tray of 5 cm × 10 cm, as described above in the instructions for the Rods in Blocks game. Alternatively, players can use the same foam frame described above in the instructions for the Rods in Blocks game.

Rules: Players start with five orange rods filling their paper trays or frames. Players take turns to throw a 6-sided die and to remove from their rods the amount that matches the throw. It is important for pupils to work in component chunks in this game, so they should not be allowed to exchange rods into single white cubes.

(Continued)

(Continued)

For example, if your first throw is 3, you must exchange one orange rod for a 3 and a 7, in order to be able to remove 3 (thus practising a complements fact). A subsequent throw of 2 will require the black rod to be exchanged for a 2 and a 5, in order to remove the 2 (thus practising a component fact). A throw of 6 on your next turn will require the 6 to be mentally partitioned into 5 and 1, and the next orange rod to be physically partitioned into 1 and 9, so that the 6 can be discarded in two chunks: 5 and 1 (thus practising decomposition as well as one of the component facts of 6). At the end of the game, the final throw need not be the exact amount (i.e. it can be more). The winner is the first player to empty his/her tray.

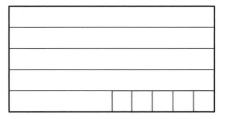

The position in the game Descent after 5 has been thrown and removed.

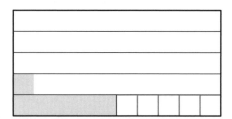

A subsequent throw of 6 would require decomposition before removal.

Variation: Start with rods to the value of 20 or 30 for a shorter game.

Card games

Play component games with cards

Sixteen card games are presented in this section. The first seven are traditional solitaire games, or adaptations of traditional patience games. They can all be played with a standard deck of playing cards, as can the eighth, ninth and tenth game.

The final six card games in this section have all been invented to target common areas of weakness in arithmetic and are best played with digit cards. Because it is difficult to find commercially produced cards in which the numbers 6 and 9 are easily distinguished, reproducible digit cards are provided on the CD. ✪

Players must not be allowed to count in ones, on their fingers or otherwise, during play. The whole purpose of getting pupils to play the games presented in this chapter is to provide them with practice in chunking, partitioning and manipulating small numbers.

Tens Away

This solitaire game practising the component pairs of 10 is a version of the traditional game sometimes known as Clear the Deck. It is given the name Tens Away when the target number is 10 (see also variations below). The game can be played with a normal pack of playing cards from which the picture cards and the 10s have been removed (an Ace counts as 1), or with a pack of digit cards made of four each of the digits 1–9 inclusive.

Rules: Deal out nine cards face up in a 3 by 3 array. Remove any pair of cards that add up to 10 and fill the empty spaces with new cards from your pack. As you remove the cards, name the complement pairs aloud, e.g. say *Seven and three is ten*. The game is won if you can deal out all the cards in the pack.

Variations: To play the traditional Clear the Deck game with a number other than 10 as the target number, remove from the pack all the cards showing the target number and all numbers higher than the target, and lay out an array of one card less than the target. For example, to practise the components of the number 7, play with 4 cards each of the numbers 1–6 inclusive, and set out an array of 6 cards to start the game.

Eleven Up

This is a traditional solitaire game for practising the components of 11. Use a pack of playing cards, counting each Ace as 1.

Rules: Shuffle the cards and set out the first nine cards, face up. Begin to deal out the remaining cards by covering any picture card (Jack, Queen or King) and also any pair of cards that add up to 11. As you play, name the two numbers that combine to make 11. The game is won if you can deal out all the cards in the pack.

Thirteens & Fifteens

This variant of a traditional solitaire game practises pairs of components of the numbers 13 and 15, with an inbuilt encouragement to construct the larger number. In this game, Kings count as 13, Queens as 12, Jacks as 11 and Aces as either 1 or 14.

Rules: Shuffle a pack of playing cards and lay out the top nine cards, face up. Remove from the array any two cards (except for Kings, see below) that add up to either of the target numbers. Put the card pairs totalling 13 in a winning pile at the left and the card pairs totalling 15 in a winning pile at the right. As soon as cards are removed from the array, fill up the empty spaces with new cards from your pack.

A King may either be removed as a single card and placed in the 13 winning pile, or matched with a 2 and placed in the 15 winning pile. Aces are matched to each other and are put in the 15 winning pile in pairs. The game is won if you can deal out all the cards from your pack and if there are more cards in the right-hand winning pile than the pile at the left, i.e. if you have accumulated more cards totalling 15 than cards totalling 13.

Card layout for Tens Away, for Eleven
Up and for Thirteens & Fifteens.

Card layout for Fifteen in a Suit
and for Standing Aces.

Fifteen in a Suit

This traditional solitaire game provides practice in the components of 15. Use a pack of playing cards from which the four 10s have been removed. Count each Ace as 1.

Rules: Shuffle the 48 cards and lay out the first 16 cards, face up. Remove any combination of cards that add up to 15, providing the cards are all of the same suit, and name the number bonds that make 15 as you match and clear the cards. Fill the empty spaces with cards from your pack. As soon as a Jack, Queen and King are all showing, you may remove all three, irrespective of their suits, and fill the empty spaces with new cards. The game is won if you can deal all the cards in your pack and clear the card layout from the table.

Variations: A harder version of this game is to remove the King, Queen and Jack only if they are of the same suit. An easier version of this game is to remove all the picture cards, together with the 10s, at the start.

Standing Aces

This solitaire game provides practice in adding pairs of numbers below 10. Use a pack of playing cards from which the Kings and Queens have been removed. Count each Jack as 11.

Rules: Shuffle the cards and lay out 16 cards, face up, in a 4 by 4 array. Aces cannot be moved or used as part of any addition. Clear the other cards from the array in pairs, but only if both cards are of the same suit and are removed from the same column or the same row of the array. As you clear the cards, you must mentally add the numbers on both cards and

(Continued)

(Continued)

announce the total of the pair. As soon as spaces appear in the array, fill the gaps with new cards from the pack. The game is won when only the four Aces remain.

Variation: Because it is quite difficult to clear the last few remaining cards, you may like to keep a score of the number of cards that remain on the table, apart from the Aces, and try to beat your record in subsequent games.

Pyramid Solitaire

This challenging solitaire game provides practice in the components of 13. Use a pack of playing cards and count each Ace as 1, a Jack as 11, a Queen as 12 and a King as 13.

Rules: Shuffle the cards and lay out the first 28 cards, face up, in a pyramid pattern as shown below, with cards overlapping from top to bottom. Set out a further 7 cards below the pyramid in a row so that none of the cards overlap or are overlapped by another. The rest of the pack is kept face down in your hand. The only cards that are in play are those face up and not overlapped by another card.

Look for pairs of cards in play that add up to 13. Remove them, and any Kings (worth 13 on their own) that are in play, from the pyramid and set them aside in a discard pile. During play, you must name the number bonds that make 13 as you match and clear the relevant cards. When no more cards can be removed, take the top card from the pile in your hand and try to match it with any face-up card in play to make a total of 13. If you cannot match the card, put it face up onto a reserve pile, from where it can remain in play as long as it is not covered by another card. When you have been through all the cards in your hand, you may turn over the reserve pile and go through these cards again, once only.

The game is won if all the 35 cards in the pyramid layout have been cleared.

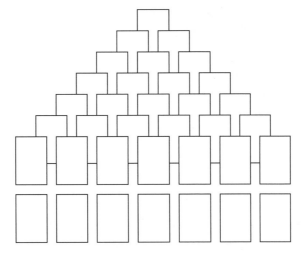

Card layout for Pyramid Solitaire

Prisoners

This solitaire game gives practice in building the numbers 11, 12 and 13. Use a pack of playing cards, from which the Aces, 2s and 10s have been removed. Count each Jack as 11, Queen as 12 and King as 13.

Rules: Take out the picture cards and place them, face up, at the top of your playing space. These 12 cards represent the 'prisoners' that must be liberated for the game to be won. Shuffle the remaining 28 number cards and deal a row of 6 cards, face up. These 6 cards are the active cards.

The release of one of the 'prisoners' can be 'bought' by putting together any two of the number cards that add up to the worth of the picture card. In other words, a Jack can be freed by two number cards with a total of 11, a Queen can be freed by a pair of cards totalling 12 and a King can be freed by a pair of cards totalling 13. As each prisoner is released, remove the picture card and the two cards paying for its release from the game. Deal new cards from the pack to fill the gaps as they appear in the row of 6 active cards. The game is won when all the prisoners have been released.

Pontoon

This traditional game, also known as Blackjack, Vingt-et-un or Twenty-one, is a game for three or more players including one who is the dealer. Players should take turns to act as the dealer.

Rules: The dealer shuffles a pack of playing cards and deals two cards to each player, face down. Players add their cards, and aim to get the highest score without exceeding 21. Aces can score either 1 or 11. The picture cards each carry a score of 10.

The dealer goes round the players in turn who must each say whether they would like another card, in which case they say 'hit me', or whether they would not like another card, in which case they say 'stick'. A player whose cards total more than 21 says 'bust' and lays the hand face up on the table. A player may ask for a fourth card on the next round, and a fifth or more on subsequent rounds, until all the players have either gone bust or decided to stick. The dealer now turns over his own cards and goes through the same process of 'hit me' or 'stick' for him/herself with the other players looking on.

If the dealer goes bust, all the players who are still in the game win a point. If the dealer sticks at 21 or less, anyone who is closer to 21 wins a point. The dealer wins a point if he/she is the closest to 21. If the dealer scores the same as another player, with no one else scoring higher, both win a point.

Zero Blackjack

Zero Blackjack is a version of the traditional game described above as Pontoon. All the same rules apply except that it is played with a pack of cards from which all the picture cards have been removed and that players are limited to a maximum of five cards each. In this version of the game, all the black cards are regarded as positive and all the red cards as negative. Players aim for a target score of zero to win the round, or win a point for being closest to zero if no player achieves the target.

Shut the Box

This traditional component game for two or three players is often sold in the form of a smart wooden box game, but can just as easily be played with either playing cards or digit cards and two 1–6 dice.

Rules: Each player sets out and controls a set of nine cards, one for each of the numbers from 1 to 9. On your turn, throw both dice, add the two numbers thrown, and turn over any card or any combination of cards from your own set of numbers to match the total, thus removing those cards from play. In the version of the game I am proposing here, you may not match the dice throw exactly unless you have no alternative. For example, if you throw 2 and 3 on the dice, you may turn over the 5 card, or the 1 and 4 cards, but may only turn over the 2 and 3 cards if neither of the other options is available. Continue to throw and to turn over cards until you no longer have cards in play that can match the dice total. Play then passes to the next player.

The winner is traditionally the player with the fewest cards still in play, i.e. still face up. However, in the version of the game I am proposing here, players score by adding the numbers on their remaining cards. The player with the lowest score is the winner. This scoring system encourages a different strategy because it gives players an incentive to use the larger numbers wherever possible.

Banking Tens

This game for two players requires a pack of 36 digit cards made up of four cards each of the numbers 1–9. The game can alternatively be played by three players with a pack of 45 or 54 cards made up of five or six cards of each digit. The game practises complements to 10 and addition using complement facts.

Rules: Players shuffle the pack and take four cards each to start. On your turn, take a card from the top of the pack to add to your hand, and try to find two cards in your hand that add up to 10. If you are able to do so, display the pair of cards to the other players, saying the two numbers aloud, and then 'bank' the pair of cards, face down on the table. As soon as you bank a pair of cards, you can have another turn and possibly bank another 10 if the new card complements a card already in your hand, and so on. When a player can bank no (more) cards, play passes to the next player.

When all the cards in the pack have been used, players count up the value of their banked cards, which should be kept in their original pairs to aid scoring. Scoring is an important part of the teaching point of this game, so players should be shown how to score by keeping a running total while stating the subtotals aloud at each step. As a first step, players can set the cards out in pairs and count up the running total of the tens: 10 … 20 … 30 … etc. Then, they must point to each card individually and announce the running total: e.g. 3 … 10 … 15 … 20 … 21 … 30, etc. The winner is the player with 'the most money in the bank', i.e. the largest value of banked cards.

During the game, cards are kept in their complement pairs, as shown here at the left.
For scoring, keep a running total, like this: 3 … 10 … 15 … 20 … 21 … 30 … 38 … 40.

(Continued)

(Continued)

Variation: Play with digit cards representing the multiples of ten up to 100 with the digits set out in a landscape format, to represent banknotes. A pack is built from four cards each of the round numbers (multiples of ten) between 10 and 90 inclusive. 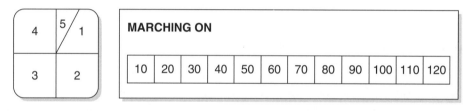 Players put money in the bank as soon as they have a pair of cards adding up to 100. Insist on the same scoring system at the end of the game as described above, so that players practise keeping a running total in which every alternate card brings the running total up to a round hundred.

Marching On

This game is for two players and requires a game board for each and an augmented set of digit cards. Start with a pack of digit cards made of four cards each of the numbers 1–9, then add in all the numbers 1–4 inclusive from a second pack, shuffling all 52 cards together. The game board is made of a track of 12 adjacent rectangles. 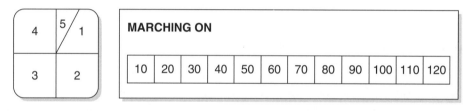 The game also requires a spinner that is divided into four equal sections labelled 1 to 4 with a small slice taken from the quarter allocated to the number 1 to make space for the number 5.

The game gives practice in building the number 10 from two or more components and also gives practice in counting in tens beyond 100, across the difficult transition points.

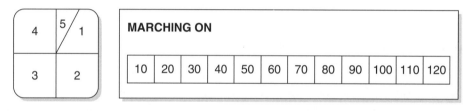

The spinner base and one player's board for the game Marching On.
Game boards can be marked with any twelve consecutive multiples of ten.

Preparation: Players each have their own game board, marked with the same set of numbers as each other. An early game might use the board shown here, on which players go 'marching on' from zero to 120; later games might start at 50, or 150, or 970, or any round number (multiple of ten) that provides a few steps of run-up before a difficult transition point.

The CD 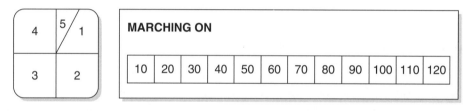 also provides a blank format for the game, to allow any suitable range of twelve consecutive multiples of ten to be selected, depending on which crossover points in the counting sequence are difficult for your particular pupils.

Rules: Each player takes it in turn to spin the spinner. On your turn, take as many cards as match the spin. If any combination of cards adds up to 10, display them to your opponent while naming them aloud, then put them face down on top of the first round number (multiple of ten) on your track blocking the round number from view. You must announce this aloud as a movement along the track, e.g. 'I can move from zero to 10', or 'I have put together another 10, so now I can go from 20 to 30'. You can go as many times as your cards allow, but only in steps of ten, before play passes to the next player.

(Continued)

(Continued)

If the cards run out before either player has reached the end, as can happen if the game is a close one, players can mark their position on the track with a token, and the cards already used can be collected, shuffled and reused. The winner is the first player to reach the end of the track.

Variation: To play a version of the game called Marching Back, all the digit cards are deemed to represent negative numbers; alternatively, make a pack of negative digit cards for the purpose. Players put together two or more cards to make a total of minus 10 for every step that they travel back from right to left, from the end of the track to the beginning.

Conjure the Number

This game for two players is played with a pack of 40 digit cards with four cards each of the numbers 1–10.

Rules: Players shuffle the pack and take four cards each to place face up in front of them. The rest of the pack is placed face down between the players. Players take turns to turn over the top card from the pack and to try to make this target number by either adding or subtracting the numbers on any two of their own cards. If successful, the player wins all three cards and takes two new cards from the pack to fill the gaps. If the player cannot make the target number, the card remains on the table and is covered by the next player's target card. If successful, the next player wins this extra card together with the three cards won on the turn.

For example, in the layout shown below, the player at the top makes the target number by subtracting 2 from 9, wins the three cards (7, 2 and 9) and replaces the 2 and 9 with new cards from the pack. The player at the bottom now turns over a card from the central pile and hopes it will show a new target of 1, 5, 6, 7 or 8.

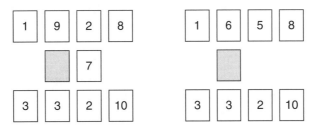

Two successive snapshots from a game of Conjure the Number.

Variation: Introduce ten extra digit cards into the pack to represent one each of the numbers 11 to 20.

Subtract from 15 (or from any other teen number)

This subtraction game for two players is played with a pack of 40 digit cards with four cards each of the numbers 1–10.

Rules: Players have five cards each throughout the game, taking new cards from the pack as soon as cards have been played.

Players take turns to lay down any card from his/her hand and challenge the opponent to subtract it from the agreed number and then match the solution with any combination of cards from his/her own hand. If successful, the second player wins all the cards on the table. If unsuccessful, the card is added to the winning pile of the player who laid it down.

For example, if your cards are 9, 9, 4, 3 and 1, you could win cards if your opponent puts down a 1, by playing 9, 4 and 1, and winning these three cards as well as the opponent's 1. Similarly, you could win if your opponent challenges you with a 2 (you would play 9 and 4), a 3 (you would play 9 and 3), a 5 (you would play 9 and 1), a 7 (you would play 4, 3 and 1) etc. However, you could win nothing if your opponent puts down a 4 or a 9. Play continues until there are insufficient cards in the pack for each of the players to keep five cards in their hand. The winner is the player who has won the most cards.

Variations: Choose any teen number other than 15 as an agreed starting point for the game.

Minimise the Difference

This subtraction game for two players is played with a pack of 40 digit cards with four cards each of the numbers 1–10.

Rules: Players take turns to pick up four cards from a shuffled pack. On turning the cards face up, the player must arrange the numbers into pairs in such a way as to minimise the difference between the two pairs of numbers. For example, a player whose cards are 3, 6, 7 and 4 will do better to choose the subtractions 7 − 6 and 4 − 3, which will score them a minimal 2, than to choose either of the other alternatives.

The player writes down the two subtraction sums to represent the working, as shown here, and scores the total of the two differences. The winner is the player with the lowest score after 3 rounds.

In the game Minimise the Difference, cards are arranged into pairs and recorded as subtractions.

Maximise the Difference

This is a subtraction game for two players, which is somewhat harder and more complex than the related game above. It is played with a pack of 40 digit cards with four cards each of the numbers 1–10.

Rules: Players take turns to pick up four cards from the shuffled pack. On turning the cards face up, the player must arrange them into pairs in such a way as to maximise the difference between two pairs of cards. For example, a player whose cards are 3, 6, 7 and 4 will do better to choose the subtractions $7 - 4$ and $6 - 3$ or $7 - 3$ and $6 - 4$, either of which would score 6, than the alternative combination of $7 - 6$ and $4 - 3$, which scores only 2. The player writes down the working as two subtraction sums, and scores the total of the two differences on that round.

After each player has played four rounds, players take their four scores and arrange these four numbers so as to maximise the difference between pairs of the numbers, in just the same way as they did when dealing with the cards. The winner is the player with the highest final score.

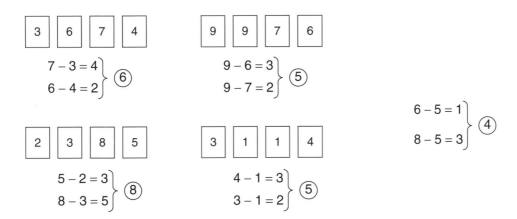

In Maximise the Difference, the scores from 4 rounds are arranged and recorded as yet more subtractions to maximise the difference between four numbers. This player's final score is 4.

Other games and puzzles

Paper and pencil puzzles that encourage component thinking

Paper and pencil games and puzzles treat numbers as abstractions. The suggestions below, therefore, are recommended only for pupils who have already been given the opportunity to learn about numbers in a concrete way through exploring and manipulating physical apparatus.

It is of the utmost importance to ensure that players are not allowed to count up in ones, on their fingers or otherwise, since this is exactly what we are trying to eradicate by encouraging pupils to engage with these puzzles and games.

Make and play a game of Plus or Minus

This game provides mixed practice of mental addition and subtraction. It also reinforces the idea that the steps within a string of additions and subtractions can be carried out in any order, i.e. that $4 + 2 - 3$ is equivalent to $4 - 3 + 2$ or $2 - 3 + 4$.

The game requires pre-prepared cards. The activity of creating the cards will provide pupils with as valuable a learning experience as actually playing the game, if you insist that all calculations are performed mentally. It is a good idea to set a limit on the size of the numbers used when creating the cards, for example by stipulating that none of the numbers can be greater than 30, or declaring that all target numbers must be 20 or less.

A template for making the cards for Plus or Minus can be found on the CD. 💿 Each card shows a number surrounded by four components. All four components must be used once, and once only, to make the target number in the middle. In this activity/game, the operations are limited to plus and minus. For example, on the first card shown below, the number 31 can be created by adding 24, 6 and 4 and subtracting 3. Try the second card for yourself.

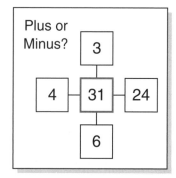 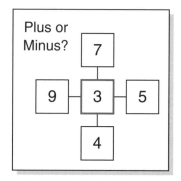

Rules: The game is played with two or three players and a small stack of puzzle cards. Either play so that each pupil has a turn to solve the puzzle within one minute before the same card is offered to the next player, or, if the pupils are evenly matched in ability, play so that each card can be won by the first person to write down a solution.

Magic Squares and Number Puzzles

A magic square is one in which all the rows, columns and diagonals add up to the same number, and where each number appears only once. Many similar puzzles are widely available, including on the Internet. Here are four examples.

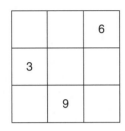

Use all the numbers from 1 to 9.
Each line must total 15.

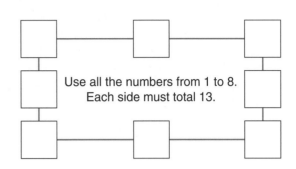

Use all the numbers from 1 to 8.
Each side must total 13.

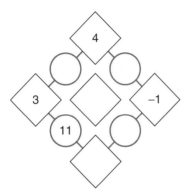

Fill in the missing numbers so that the numbers in the circles are the sum of the numbers in the two diamonds on either side.
Add the four corner numbers to complete the central diamond.

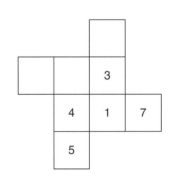

Fill in the missing numbers so that each 3-cell column and 3-cell row has the same total.

Component Su Doku

In the Su Doku puzzles shown below, the numbers from 1 to 5 can appear only once in each row and in each column. The thicker lines enclose two different components that add up to the number in the top left corner of the enclosure.

The two component Su Doku puzzles shown here, together with another 22 Su Doku component puzzles at different levels of difficulty, can be printed off the CD. 🔘 The puzzles at the moderate level can be solved by a process of elimination together with an understanding of the limited possibilities for splitting very small numbers into components. To solve the intermediate puzzles, pupils must also be able to work out the total values in each row and column and use that fact to make deductions. The most difficult puzzles require several different techniques and the versatility to choose the appropriate strategy at each stage of the solution. All the puzzles provide excellent opportunities to practise building up and breaking down numbers into components.

(Continued)

(Continued)

If you find your pupils trying to solve these puzzles by counting up on their fingers, give them more practice in some of the earlier component games using Cuisenaire rods, such as Who Has the Last Word?, or Descent, or Rods in Blocks. Alternatively, provide the pupils with a set of Cuisenaire rods and insist that they build all the numbers below 8 out of two different components, before tackling the Su Doku puzzles. The rods should stay within view so that pupils can use them for reference. Pupils will be able to see from the rods that there is only one acceptable way of building the numbers 3 and 4 (since the doubles combination of 2 + 2 would break the Su Doku rule of different numbers within each row and column); that the only possible combinations for building 5 are 1 + 4 or 2 + 3; that the only two acceptable ways of building 6 is by pairing 1 + 5 or 2 + 4; that there are only 3 possible ways of building 7 from two components, etc. Discovering how very few combinations are possible often reassures pupils who might otherwise believe that they are expected to memorise an unfeasible number of component facts.

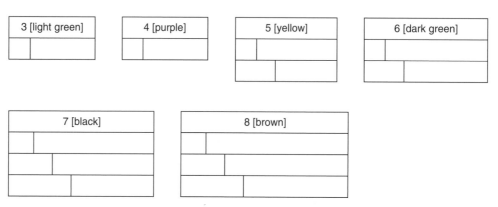

Building small numbers out of two different rods highlights how few possibilities there are, thereby reassuring pupils that there are not too many component facts to learn.

These Su Doku puzzles should always be solved by logic and reasoning, not by guesswork or trial and error.

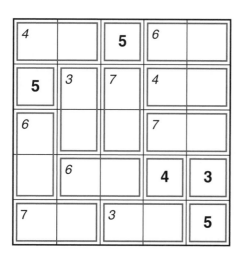

In these Component Su Doku puzzles, the numbers from 1 to 5 appear once in each row and in each column. The thicker lines enclose two different components that add up to the number in the top left corner of the enclosure.

(Continued)

(Continued)

Pupils who enjoy the Su Doku puzzles shown here and on the CD 💿 might also like to try the mild versions of the 'Killer Su Doku' puzzles that are published in some daily newspapers.

A variant of the Component Su Doku puzzle is the Difference Su Doku puzzle, in which clues are given about the difference between two components, i.e. the subtraction fact. Two examples are shown below.

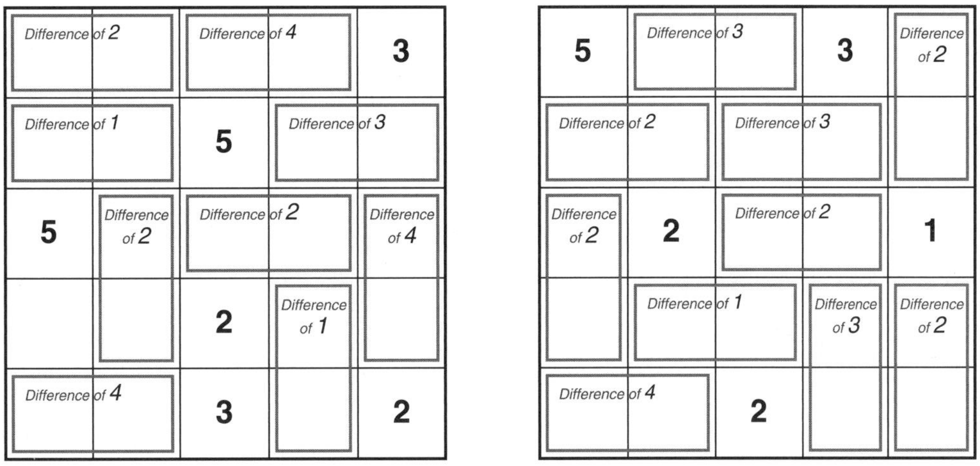

In these Difference Su Doku puzzles the numbers from 1 to 5 appear once in each row and in each column. The difference between the two components within a single enclosure is noted at the top of the enclosure.

References

1 E. Gray (1997) 'Compressing the counting process: developing a flexible interpretation of symbols', ch. 6 in *Teaching & Learning Early Number*, ed. I. Thompson, Open University Press.
2 R. Bird (2007) *The Dyscalculia Toolkit*, Sage.
3 B. Butterworth and D. Yeo (2004) *Dyscalculia Guidance*, nferNelson.
4 M. Sharma (various 1980–93) *Math Notebook*, Center for Teaching & Learning of Mathematics, Framingham, MA, USA. (Professor Sharma's publications and videos are available in the UK from Berkshire Mathematics: www.berkshiremathematics.com)

PART II

The bridging technique

CHAPTER TWO

Pre-skills for learning the bridging technique

Overview

Bridging through 10, or through multiples of 10, is the single most useful mental calculation strategy that pupils can learn.

Bridging through 10 is the technique by which addition is performed as a linear movement on a number line. Pupils must be introduced carefully to the idea of a number line and shown how it differs from a number track. The number line on which bridging is modelled will, at first, be an actual line or a sketch of a line, but with sufficient practice pupils can learn to visualise an imaginary number line on which to perform mental addition. Crucially, movement along the line is not performed one step at time, but in just two jumps. The number 10, a significant number in our decimal number system, acts as a stepping stone between the two jumps. The other important and notable feature of bridging is that work can always be performed from left to right, in the forward direction.

Before learning the bridging technique, pupils must first have mastered certain pre-skills. I intend the word 'master' to convey a state in which concepts have been thoroughly understood and internalised. Ideally, the necessary pre-skills will have been acquired at primary school, but for older learners who still need to work on these skills and concepts, teaching suggestions can be found in this chapter. For learners who have already mastered the pre-skills, turn to Chapters 3 and 4 for a step-by-step guide to teaching the bridging technique.

Summary of necessary pre-skills for the bridging technique

1. Complements to 10.

2. Components of the whole numbers up to 10.

3. The connection between addition and subtraction.

4. The difference between a number track and a number line.

(Continued)

(Continued)

5. Subtraction as complementary addition.

6. Reasoning from known facts.

7. Adding a 1-digit number to 10.

8. Understanding the decimal structure of the number system.

9. Place value basics.

10. Building and partitioning 2-digit numbers.

In the lists of teaching suggestions below, asterisks denote those activities or games that are described in detail in my book *The Dyscalculia Toolkit*[1] and that are therefore not spelt out again here.

Necessary pre-skills for bridging through 10

1. Complements to 10

The pairs of numbers that add up to 10 must be known by heart. I use the word 'complements' with my pupils, and throughout this book, as a convenient label for component pairs of numbers adding up to 10, or to other round numbers (multiples of 10). Pupils are relieved to find that there are only five or six complement facts to learn by heart: $5 + 5$, $4 + 6$, $3 + 7$, $2 + 8$, $1 + 9$, $(0 + 10)$. A large number of teaching activities and games are presented in *The Dyscalculia Toolkit* so that teachers can revisit this important topic again and again while offering plenty of variety to maintain pupils' interest.

Some suggested teaching activities are:

▶ Bead string activities.*

▶ A Cuisenaire staircase extended to make a wall.*

▶ Complement number searches.*

▶ Complement ping-pong game.*

▶ Pelmanism card game.

▶ Ten in a Bed game.*

▶ Clear the Deck solitaire game.*

▶ Some of the domino games and other component games from Chapter 1.

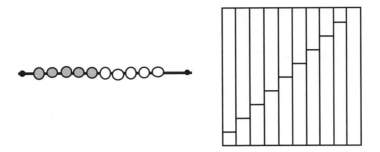

A bead string made of two colours, and Cuisenaire rods extended from a staircase into a wall, are useful aids for learning the complements to 10.

2. Components of the whole numbers up to 10

Although the components of the number 10 are the most important, pupils must also recognise the components of all the other whole numbers below 10. Early understanding is fostered by building number patterns from discrete objects, such as counters or nuggets, and manipulating the objects so that quantities below 10 are physically partitioned and recombined. Dice patterns, and Dorian Yeo's dot patterns,[2] are particularly suitable for young pupils. I find that older pupils often prefer to work with domino patterns (see Chapter 1).

Pupils should not be required to remember all the number bonds to 10 (except the five complement pairs), but should instead be trained to visualise a number by focusing on a pattern so that they can 'see' and manipulate its component parts in their mind's eye.

Some suggested teaching activities are:

▶ Domino games (see Chapter 1).

▶ Individual numbers built and explored with Cuisenaire rods.*

▶ Cuisenaire rod sandwiches*, including those with missing rods.*

▶ Component games with Cuisenaire rods or cards (see Chapter 1).

▶ Shut the Box/Cover the Number game.*

▶ Clear the Deck solitaire game.*

▶ Component Su Doku puzzles (see Chapter 1 and the CD).

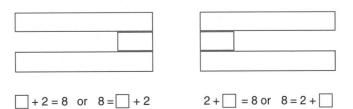

$\square + 2 = 8$ or $8 = \square + 2$ $2 + \square = 8$ or $8 = 2 + \square$

Cuisenaire rods sandwiches can be set up with a missing rod to explore components of a number. These rods should be read as 'What must be added to 2 to make 8?' before being written as a missing-addend problem, with the missing number in either position.

3. The connection between addition and subtraction

Addition and subtraction should be taught alongside each other in ways that reinforce the underlying connection between them. Pupils should be shown how both addition and subtraction can be used to describe or record a single relationship, from different points of view. I sometimes use the analogy of a brother and sister: the girl says she has a brother, the boy claims a sister, but both are describing the same relationship, each from a different perspective.

Some suggested teaching activities are:

▌ Missing rods from Cuisenaire rods sandwiches.*

▌ Equations made of Cuisenaire rods to build, manipulate and read.*

▌ Simple equations to record on paper as both additions and as subtractions.*

▌ Hidden quantity scenarios recorded in various ways.*

▌ Word problems to be made up by pupils to match a given number fact.*

▌ An informal triangular notation of the triad of related facts. For example, using diagonal arrows or lines to show that 8 can be made of 2 and 6.

An informal triangular notation for related addition/subtraction facts.

4. The difference between a number track and a number line

These two terms are not interchangeable: they describe two different concepts. Pupils who confuse the two often produce answers that are one more or less than the correct solution.

A number track is a model of the number system that allocates a defined space, or area, to each number. Pupils' first encounter with a concrete number track at school might be an activity that involves putting counters into a line and touching each one as they count to find the total amount. Before school age, many children will already have become familiar with diagrammatic number tracks in the form of board games, such as Snakes & Ladders. However, the number track is not a model that is exclusively associated with young children and small numbers, as can be seen by the 100-square that is widely used in primary and middle schools. In order to calculate on any version of a number track, pupils count actual numbers, and usually do so by reckoning in ones.

A number line, by contrast, is a much more abstract model. Whether the line in question is an empty number line drawn on paper or an actual line numbered in ones or tens along its whole length, for example the edge of a ruler, what is being recorded or measured is the abstract idea of the intervals between numbers. In order to calculate successfully on a number line, a pupil does not count actual numbers but focuses on the steps or spaces between them. Making no clear distinction between number tracks and number lines leads many children, and especially dyslexic and dyscalculic learners, to be permanently confused about whether they are supposed to start a calculation with the number that is given or with the next number. The fact that empty number lines model intervals between numbers is what makes them ideal for adding and subtracting in chunks, rather than in ones.

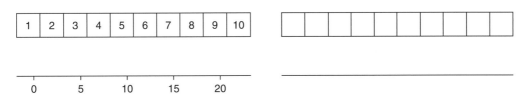

Number tracks (top) are constructed from areas, with one space for each consecutive number.
Number lines (bottom) show the intervals between numbers. Labelled numbers are not necessarily consecutive.

Some suggested teaching activities are:

▌ Draw Your Race on a Number Line game.* Adapt the game for older learners by using a track of fifteen to twenty consecutive 2-digit numbers.

▌ Take care to use the correct terminology. Explain explicitly to pupils the difference between a number track and a number line.

▌ Work from a number track to an empty number line: pupils build a 2-digit number by placing Cuisenaire rods onto a 1–50 track made of 1 cm squares. Propose a single-digit addition requiring no bridging. Pupils record the numbers, the calculation and the solution on an empty number line.

▌ Work from a number line to a number track: make up a calculation, without the solution (and with no bridging required) and sketch it on an empty number line. Pupils match it concretely with rods on a track.

▌ A paper 1–100 number track cut into strips of ten numbers which are then stacked one under the other to create a 100-square.* The resulting square is used to show the complement of any 2-digit number to the next multiple of ten, and to find the complement to 100, after which both complement facts should be recorded by pupils on an empty number line.

5. Subtraction as complementary addition

Pupils with dyscalculia or other specific difficulties in maths normally find it extremely difficult to work backwards. Such pupils should be taught to solve both addition and subtraction problems by working in the forward direction. Subtraction can be understood as complementary addition, by choosing to take away the quantity to be subtracted from the *start* of a counted quantity, or from the beginning of a number line. The solution is reached by finding the difference, or the gap, between the two numbers. Dorian Yeo's idea[3] of scribbling out part of the number line, to represent the amount that is being taken away, is something that many pupils find helpful when they begin to use empty number lines for solving and recording complementary addition.

For example, to model the subtraction 17 – 13, draw a length of line to represent 17, and show 13 being taken away from the beginning of the line, the section representing zero to thirteen inclusive, so that the answer can be seen to lie in the remaining gap between the numbers 13 and 17.

Example: 17 – 13

Some suggested teaching activities are:

▶ Complementary addition demonstrated with counters or nuggets.*

▶ Complementary addition on number lines beginning at zero.*

▶ Complementary addition on empty number lines.*

▶ Subtraction reframed as 'finding the difference'.*

▶ Subtraction problems rewritten by pupils as equalising or missing-addend problems, e.g. 17 – 13 = ☐ also means 13 + ☐ = 17.

▶ Complementary addition as a way to avoid the difficult decomposition demanded by some column subtraction problems.*

▶ Investigation of the language of word problems matched to a subtraction and to the related missing-addend sum. For example, for the problem 15 – 7, a suitable word problem might be: *How many coins remain after Sam spends seven of the fifteen coins he had in his pocket?* The related missing-addend problem might be: *How many more coins would Sam need to add to the seven coins in his pocket in order to have fifteen coins?*

▶ Money and measurement problems involving complements to 100.*

6. Reasoning from known facts

Pupils must be taught explicitly how to derive new facts from those they already know. At this level, the pre-skill level, our aim is to introduce pupils to the idea of using logic to extend their knowledge of number facts. For example, after building a number from two groups of concrete materials, such as counters or nuggets – i.e. discrete materials – demonstrate how the process of moving an object from one group to the other must result in one group becoming smaller by one while the other group becomes larger by one. Talk through the process and encourage pupils also to verbalise what they see. When examining two neighbouring rows within a sandwich made of Cuisenaire rods – i.e. continuous materials – point out that when one component becomes larger by one unit, the other component must automatically become smaller by one unit, if the total is to remain the same. For example, if 4 and 3 make 7, it follows that 5 and 2 also make 7.

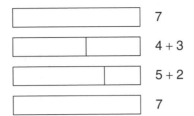

Later, have the pupils practise reasoning from complements facts and doubles facts. For example, if 4 + 6 = 10 (a known complement fact) it follows that 4 + 7 = 11 or 4 + 5 = 9, or 5 + 6 = 11 or 3 + 6 = 9. If 7 + 7 = 14 (a known doubles fact) it follows that 6 + 7 = 13 or 7 + 8 = 15, etc. For much more on reasoning strategies, see Chapter 9.

Some suggested teaching activities are:

▶ Numbers built out of two components using counters or unit cubes. Addition sums are written down to record what happens when one item is moved from one group to the other, e.g. 2 + 6 = 3 + 5.

▶ Numbers built out of two lengths of Cuisenaire rods placed end to end. Pupils are asked to identify all the number bonds that can be derived by one step of reasoning from the fact modelled by the rods. The newly derived facts should be expressed as subtractions as well as additions. For example, a purple and yellow rod placed end to end can be read as 4 + 5 = 9 or as 5 + 4 = 9 or as 9 = 5 + 4 or as 9 = 4 + 5 or as 9 − 4 = 5 or as 9 − 5 = 4. One step of reasoning from these facts can produce just as many different ways to express each of these combinations: 3 + 5, 5 + 5, 4 + 4 and 4 + 6.

▶ Mixed sums presented in writing, of which many can be solved by near-complements facts and near-doubles facts, i.e. sums for which the answers can be derived from knowledge of complement pairs and of doubles.* Pupils score points only for correctly answering the target questions, but lose marks for answering any others, to ensure that pupils are focusing on the strategy and not just on the answer.

▶ Ten difficult additions written on the board, complete with answers that have been provided by pupils using a calculator. Pupils then answer related questions without a calculator by reasoning from one of the facts on display. For example, if one of the sums on the board is 28 + 56 = 84, pupils can be asked, *What is 56 and 29?* or *What must be added to 38 to make 84?* or *What is 84 minus 57?* etc.

▶ Sums based on known doubles facts made up by pupils for each other to solve, e.g. 8 + 7 or 25 + 26.

7. Adding a 1-digit number to 10

This is best understood by building numbers out of Cuisenaire rods in the most efficient way. For example, to build the number 12, an orange rod and a red rod are put together, showing that 12 is made of 10 and 2, or one ten and two units. All the numbers between 10 and 20 should be built, often, in a random order, and then recorded as sums, e.g. 12 = 10 + 2 or 10 + 2 = 12.

Note that the teen numbers can cause difficulty because they are irregular: the names for 11 and 12 hide the fact that they are teen numbers at all, and the remaining teen numbers have a reversed verbal construction in which one hears and says the part of the number name representing the units digit in advance of the suffix '-teen' (akin to working from right to left), at the same time as having to read or write the digit representing the ten before the units digit (working from left to right).

Some suggested teaching activities are:

▶ All the teen numbers built, in random order, on place value mats using Cuisenaire rods or Dienes blocks. Once built, the amounts are read as numbers and recorded in writing in digits.*

▶ All the teen numbers built from 10p and 1p coins, or 10 cent and 1 cent coins. Once built, the amounts are read and recorded in writing in digits.

▶ Number tracks that are 20 units long, on which pupils find the position of a number and the position of the number that is ten more or ten less, without counting on or back in ones.

▶ A paper number track cut into strips of ten numbers which are then stacked one under the other to create a 100-square.* The square is used to find any number chosen at random and then immediately to find the number that is ten more or ten less.

▶ The Jump 10 game.*

▶ Random numbers, generated by a 0–9 die or spinner, to which pupils add 10 mentally, before recording the addition on an empty number line.

Example: 8 + 10

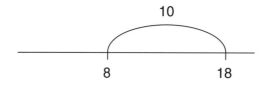

Adding 10 does NOT require bridging.

Further necessary pre-skills for bridging through multiples of 10

8. Understanding the decimal structure of the number system

By the time pupils reach secondary school they must have a reasonable understanding of the place value system in order to make sense of the maths curriculum. Many computational errors can be traced back to muddled or distorted ideas about place value. See also pre-skill 9, below.

The teen numbers present a problem to some learners who struggle with maths because the number names are not regular, as mentioned above. This irregularity, which most pupils meet early in their schooling but which usually remains unacknowledged, can serve to mask the strong decimal structure of our number system. For example, because the names for the numbers 11, 21 and 101 have only a weak auditory relationship, dyscalculic pupils may not realise that the structure of the numbers themselves exemplify a close relationship, since all are built from one/some tens and a single unit.

Creating a 100-square by cutting up and reassembling a number track is one of the best exercises to help pupils notice and internalise the way the pattern of the first ten numbers is repeated in every subsequent decade of numbers.

Pupils should not be expected to work with larger numbers until they are thoroughly familiar with all the numbers up to 100 and can put a 2-digit number into context, for example by knowing which two multiples of ten lie on either side. Furthermore, teachers should not assume that an understanding of the numbers up to 100 means that pupils will be able to transfer their understanding to numbers up to 1000 automatically: pupils with dyscalculia or other specific maths difficulties have a very poor sense of the magnitude of numbers as well as a muddled idea of the sequences that make up the decimal number system, and will need lots of practice with concrete materials to improve their feel for numbers.

Some suggested teaching activities are:

▶ An extended staircase of 20 steps made from Cuisenaire rods.*

▶ 2-digit numbers built from rods or base-ten materials on place value mats and recorded in writing in digits in columns.*

▶ Random 2-digit numbers to be found as quickly as possible on a number track. Later pupils can be asked to find the same numbers on an almost blank track, on which only the multiples of 10 are labelled and the halfway points between them.*

▶ A paper number track cut into strips of ten numbers which are then stacked one under the other to create a 100-square.* Pupils use the created square to find any randomly chosen 2-digit number and to name and find the nearest multiple of ten that follows the target number.*

▶ An empty number line on which to record the distance between any 2-digit number and the two multiples of ten on either side.*

▶ The Magic 10s game.*

▶ Win Counters on a 100-Square game.*

▶ Race through a 100-Square game.*

9. Place value basics

As I explain in detail in *The Dyscalculia Toolkit*, many pupils who struggle with maths are confused by the abstract idea of place value. Those who have not grasped that place value column names have a logical and repeating pattern do not really understand numbers with 4 or more digits.

The difficulty seems to arise because teachers sometimes assume that pupils will notice patterns within a bigger picture and therefore do not think to draw pupils' attention to these patterns explicitly. Most children are first introduced to single-digit numbers, and, as they get older, to 2-digit numbers (which we teach them to call tens and units) and then to 3-digit numbers (introducing the hundreds) and later still to 4-digit numbers (introducing thousands). This gradual introduction leads many children to assume that every column in our place value system has a new name. This is a very common misconception.

In fact, completely new labels are given not to every column, but to every group of THREE columns. Each threesome belongs to a 'family' that is labelled according to its place value magnitude – ones, thousands, millions, billions, etc. Thus, multi-digit numbers have (starting at the right): units, tens and hundreds of *ones*, then units, tens and hundreds of *thousands*, then units, tens and hundreds of *millions*, then units, tens and hundreds of *billions*, etc.

Millions			Thousands			Ones		
Hundreds	Ten	Units	Hundreds	Tens	Units	Hundreds	Tens	Units

The three fold repeating pattern of Hundreds, Tens and Units columns.

Some suggested teaching activities are:

▶ An extended staircase from 1 to 20 made of Cuisenaire rods.* Individual columns are removed from the staircase and arranged on place value mats. Pupils read aloud and write in digits the number modelled by the concrete materials.

▶ A written 2-digit number to be matched by pupils using base-ten material or Cuisenaire rods, and also shown on a Slavonic abacus and on a 3-spike abacus. *N.B.* if you choose to work on a spike abacus, it is best to have 3 or 6 spikes showing, never 4 or 5 spikes only, so as to reinforce the threefold nature of place value column names.

▶ An empty number line used by pupils to record the distance between any randomly chosen 2-digit number and the two multiples of ten on either side.* Pupils can extend this activity by using the same number line to mark the two multiples of 100 on either side in order to put the target number into a wider context (but without attempting to calculate the distances between the numbers at this stage). The whole exercise can be usefully repeated with numbers of 3 or more digits, while still focusing on the multiples of 100 on either side of the target number.

▶ Familiar additions and subtractions, modelled on renamed Cuisenaire rods and recorded on paper both horizontally and in columns. For example, take a red and a purple rod that might normally be recorded as $2 + 4 = 6$, $6 - 4 = 2$, etc., and challenge pupils to record the relationships if the rods were now named 20 and 40, or 200 and 400, or 2000 and 4000.

▶ A 3-spike abacus and an empty number line, used side by side, to model additions and subtractions of either multiples of 10 or of 100. Only examples that do not require carrying or decomposition should be chosen, such as $56 - 10$ or $145 + 200$. Pupils write the problem in digits, both horizontally and vertically, and highlight the single digit that has been altered, using a second colour to highlight the remaining digit or digits that have not been affected by the calculation.

▶ Explicit teaching of the threefold repeating pattern of columns in the written number system, and how to read and write multi-digit numbers.*

▶ Place value notation presented as a time-saving shorthand, compared with the alternative of having to write multi-digit numbers using words to spell out the values of each digit.*

▶ Dice and spinner games in which amounts are built concretely and then altered during play.*

▶ The Four Throws game.*

▶ Steer the Number (an Ian Sugarman game[4]).*

10. Building and partitioning 2-digit numbers

Many of the pre-skills described above involve building numbers concretely. It is also important for pupils to learn to partition numbers, and to learn that there are various ways of splitting numbers into useful chunks. For example, understanding how a number can be split into a teen number and the remaining group of tens is essential preparation for the kind of column subtraction that requires decomposition. As always, the best way to explore this concept is with concrete materials that pupils can manipulate freely. Abstract paper and pencil exercises are unlikely to result in true understanding.

Some suggested teaching activities are:

▶ Numbers partitioned into tens and units in various ways, using concrete materials, before recording the partitions as diagrams or in digits.* For example, 56 split into $50 + 6$, $40 + 16$, $30 + 26$, $20 + 36$, $10 + 46$.

▶ Numbers partitioned so that the teen number is split off.* For example, 56 split into 16 + 40, or 88 split into 70 + 18. Numbers should be modelled concretely with Cuisenaire rods before exposing pupils to abstract work.

▶ Numbers recorded as an addition of tens and units, e.g. 29 = 20 + 9, and also as an addition of multiples of tens and multiples of units, e.g. 29 = (2 × 10) + (9 × 1). At first, the numbers should be modelled concretely.

▶ Dice and spinner games in which amounts are built concretely and then reduced during play.*

▶ Steer the Number game (an Ian Sugarman game[4]).*

▶ The Place Value Boxes game.*

▶ The Calculator Skittles game.*

References

1 R. Bird (2007) *The Dyscalculia Toolkit*, Sage.
2 D. Yeo (2003) *Dyslexia, Dyspraxia & Mathematics*, p. 102, Whurr.
3 Ibid., p. 224.
4 I. Sugarman (1997) 'Teaching for Strategies', *Teaching & Learning Early Number*, ed. I. Thompson, p. 151, Open University Press.

CHAPTER THREE
Bridging through 10

Overview

Bridging is the single most useful mental calculation strategy that pupils can learn.

Bridging through 10 is a technique based on a linear understanding of the number system. Addition is performed by movement along a number line and by adding components in convenient chunks. The number line can, at first, be an actual line, and later a sketch of a line on paper. Later still, and only after sufficient practice, pupils should be able to visualise an imaginary number line on which to perform mental addition. The crucially important feature of the bridging technique is that the addition is not performed one step at a time by counting in ones, but is performed in just two jumps with the number 10 functioning as the stepping stone between the two jumps.

My own conviction about teaching bridging on a number line leads me to incorporate a further crucial feature: movement along the line is always in the forward direction.

Although bridging is an abstract concept, it should be introduced concretely by modelling addition with materials that are continuous rather than discrete, namely Cuisenaire rods, before pupils learn to record the work with paper and pencil on an empty number line.

An empty number line is one on which no numbers are labelled in advance. Pupils mark only what is necessary on the empty line as they work through the process of calculating a solution. I do not follow the advice found in the Mathematics and Numeracy Frameworks, which maintains that each number must be preceded by a positive or negative sign and that each jump must show the direction of movement. Because pupils with specific maths difficulties find it virtually impossible to work backwards, I teach all my pupils to work only in the forward direction. On a number line, this means that calculations are always recorded and read from left to right, removing the need for either signs or arrows.

The bridging technique should not be confined to addition problems. Bridging is particularly useful as a way of subtracting by means of complementary addition. Many teachers assume that, because addition can be thought of as movement from left to right along a number line, it is only logical to view subtraction as a movement in the opposite direction. However, most pupils who struggle with maths simply find it too difficult to work backwards, and should never be forced to do so. If pupils can manage one or two small backwards steps mentally, for a calculation such as

26 − 6, or by extension 26 − 8, I would encourage them to continue practising this useful skill. But a backward movement of more than two steps puts too great a burden on the working memory of most pupils with specific maths difficulties. I therefore teach all pupils to calculate and record subtraction on a number line from left to right, treating subtraction as complementary addition.

A significant benefit of using bridging techniques combined with the cognitive model of an empty number line is that both addition and subtraction are performed and modelled in the same way. The only difference between the two is the location of the answer. The answer to an addition question lies in the position that has been reached on the number line; the answer to a subtraction question lies in the amount of movement along the line, i.e. in the sum of the jumps.

Before learning how to bridge, pupils must have mastered the first seven pre-skills discussed in detail in the previous chapter. This chapter shows how to teach the bridging through 10 strategy one small step at a time through a series of sequential activities. The next chapter shows how to extend the teaching to bridging through multiples of 10.

Summary of activities for learning to bridge through 10

Activity 1.   Add two 1-digit numbers. Keep the second addend the same; for example, add 6 to various numbers.

Activity 2.   Add two 1-digit numbers. Keep the first addend the same; for example, add various numbers to 8.

Activity 3.   Demonstrate that addition is commutative.

Activity 4.   Model single-digit addition on an empty number line.

Activity 5.   Use number lines for solving simple subtractions by complementary addition, keeping all the numbers below 20 and the solutions below 10.

Activity 6.   Practise visualising a number line for mental calculations.

Bridging through 10

Before the first activity

Ensure the necessary pre-skills are in place (see Chapter 2).

Activity 1

Add two 1-digit numbers. Keep the second addend the same; for example, add 6 to various numbers.

Start with the number 6, say, and have the pupils make a rods sandwich to explore all the component pairs that add up to 6.

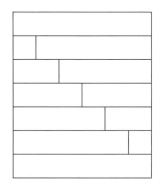

6 [dark green]

A 6 sandwich showing all the component pairs in order.

Keep the sandwich in full view when adding 6 to different numbers using rods, making sure that the examples you choose will require bridging (i.e. add 6 to a number greater than 4).

For example, for 8 + 6, take a brown rod and a dark green rod and put them end to end.

Example: 8 + 6

| 8 [brown] | 6 [dark green] |

The result is longer than one orange rod but shorter than two orange rods, as can be seen by lining up one or two orange rods alongside the brown. This tells us that bridging will be required and that the answer will be in the teens.

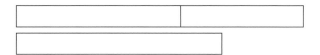

Looking at the rods, pupils can see that, in order to bridge through 10, the 6 must be split into 2 and 4, because 2 *is the complement* of the first number, 8; and 4 is *what is left* of the 6 after establishing that the first component must be 2.

Pupils can measure the two smaller rods alongside the dark green rod to prove to themselves that the rods have the same value, or can refer back to the 6 sandwich they made at the start. Pupils substitute the red and purple rods for the dark green.

I usually leave the orange rod in place when working through these kinds of examples, to emphasise the fact that we are bridging through ten; however, the orange rod can be removed as soon as the exchange has been decided upon, if pupils find its presence confusing.

(Continued)

(Continued)

Example: 8 + 6

8 [brown]	6 [dark green]

8 [brown]	6 [dark green]

10 [orange]

8 [brown]	2 [red]	4 [purple]

10 [orange]

Pupils can find the presence of the orange rod confusing when they are not clear that is being used for measuring purposes only, and therefore try to include it in the calculation. For such pupils, you may need to start this concrete work on bridging by drawing a track for the pupils to work on. The track must be 1 cm high, to match the dimensions of Cuisenaire rods, and should be at least 20 cm wide with a closed end at the left and an open end at the right. Have the pupils draw a strong line after the first 10 cm from the left, to represent the point through which we are bridging, i.e. the number 10. Pupils lay the rods they are working with on top of the track, and use their thick line for determining how the second addend, the 6 in this example, is to be partitioned.

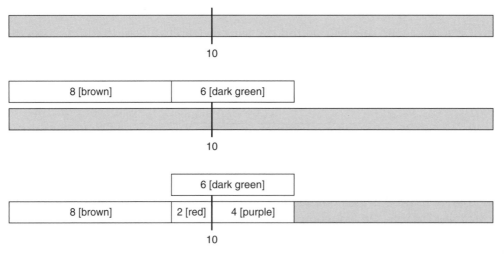

Refer back to the sandwich, and ask pupils to point to the component pairs that provide the necessary split in each case, namely, *the complement* (of the first number) *and what's left*. In this example, it is the row of the sandwich filling that shows red + purple, i.e. 2 + 4.

Point out to pupils that our focus is on the second number because that is the one that needs to be split. It is the position of the number as the second addend in the question that identifies it as the one to partition. This is a fact that has to be emphasised clearly and often.

Repeat the whole procedure with various numbers to which 6 is added. Refer back to the sandwich each time and have pupils identify the particular way that the 6 is being partitioned in each example.

(Continued)

(Continued)

Example: 5 + 6

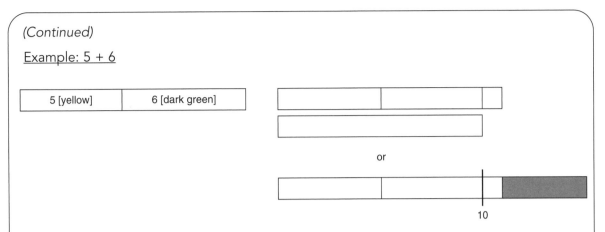

Ask the pupil to find additions in which 6 will not need to be split. [Answer: 1 + 6 ... 4 + 6.]

To test pupils' understanding, you can ask them to make up the addition question that would require 6 to be split a certain way. For example: *Can you give me a sum, in the form □ + 6, that would make me have to split the 6 into 5 and 1 for bridging?* Pupils have to deduce that if the first component of 6 is 5, and as the first component represents the complement of the first number in the sum, the problem sum has to be 5 + 6. This is, in fact, the example shown above that pupils will have already practised. Not every pupil is able to reason in this way during the early stages of working with bridging, but you should persevere with this type of question from time to time. Once pupils can answer this kind of question, you can be sure that they have really understood an important concept.

Repeat the whole of this activity for addends other than 6.

Activity 2

Add two 1-digit numbers. Keep the first addend the same; for example, add various numbers to 8.

Use Cuisenaire rods to show how the first number, say 8, remains intact while the second number is split into components. However, emphasise and re-emphasise that it is the first number that dictates how the second amount must be split, just as pupils practised in Activity 1 above. The split is always: *the complement, and what's left.*

For example, for 8 + 5, take a brown rod and a yellow rod and place them end to end.

Example: 8 + 5

8 [brown]	5 [yellow]

The result is longer than one orange rod but shorter than two orange rods, as can be seen by lining up an orange alongside the brown, or by laying the rods onto the centimetre-based track introduced during the previous activity. It can now be seen that the 5 should be split into 2 and 3: 2 is *the complement* of the first number, 8; and 3 is *what is left* of the 5 after establishing that the first component is 2.

Pupils can measure the two smaller rods alongside the yellow rod to prove to themselves that the rods have the same value, before substituting the red and light green rods for the yellow rod.

(Continued)

(Continued)

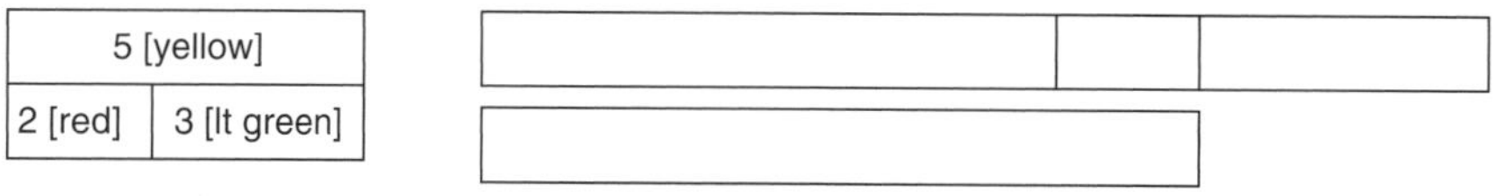

5 [yellow]	
2 [red]	3 [lt green]

Pupils should now read the rods as a number sum that bridges through ten: *8 plus 5 is the same as 8 plus 2 plus 3, or 10 and 3, which is 13.*

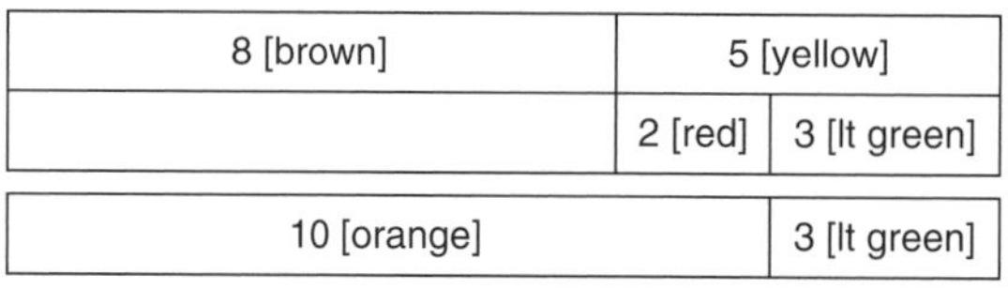

8 [brown]	5 [yellow]	
	2 [red]	3 [lt green]
10 [orange]		3 [lt green]

Repeat this activity with different numbers added to 8, not in any set order, for example 8 + 3, 8 + 6, 8 + 4, 8 + 9, 8 + 8, 8 + 5, 8 + 7.

Discuss why all the above examples require bridging, but 8 + 1 does not. Have pupils articulate what they now understand about the term 'bridging through 10'.

Repeat the whole of this activity with numbers other than 8.

Activity 3

Demonstrate that addition is commutative.

Use Cuisenaire rods to demonstrate to pupils that, no matter which number is taken first, the answer will always be the same. During this demonstration, point out to pupils that if both addends are single-digit numbers, it is just as easy to start with the smaller number as it is to start with the larger number.

For example, 5 + 8 is solved through the same process and requires the same number of steps as 8 + 5, as shown below.

Example: 5 + 8 = 8 + 5

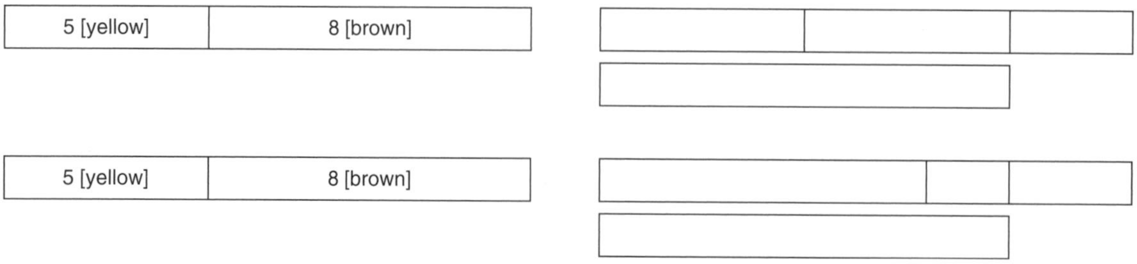

5 [yellow]	8 [brown]

5 [yellow]	8 [brown]

In each case, it is the first number that dictates how the second number must be split. The second amount is split into *the complement* (of the first number) *and what's left*.

Have pupils repeat the whole of this activity for themselves with various additions of two 1-digit numbers that add up to more than ten.

Activity 4

Model single-digit addition on an empty number line.

Two acceptable, and slightly different, ways of recording addition on a number line are shown here. Pupils should be allowed to try out both ways before choosing and sticking to one of the methods. In order to keep the sketches as clear and uncluttered as possible, discourage pupils from including positive or negative signs, and do not attach arrows to the jumps.

Both methods are shown here on the example 8 + 6. Both record the second addend being split into two – and only two – components. The method shown below at the left involves drawing two separate jumps for the number 6, to show that it is being partitioned into 2 and 4 so as to bridge through 10. The disadvantage of this method is that after the first jump, pupils must look back at the question to check that they have not yet completed the calculation and to determine what the next step should be. Pupils with specific maths difficulties and/or weak working memories often lose track of what they are supposed to be doing at this point.

The method shown below at the right involves drawing a single jump to represent the number 6, a jump that is later subdivided it into two smaller jumps that bridge through 10. The advantage of this method is that the whole problem is transferred at once to the number line, obviating the need for the pupil to look back at the original written question; the disadvantage is that it can look cramped and messy if the pupil leaves insufficient space for all the necessary workings within the first jump.

Example: 8 + 6

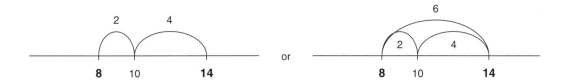

Pupils should practise their chosen method on many calculations that they have previously modelled concretely with rods in Activities 1 and 2 above.

Activity 5

Use number lines for solving simple subtractions by complementary addition, keeping all the numbers below 20 and the solutions below 10.

Pupils need to understand that a subtraction such as 15 − 7 can be solved by finding 7 + □ = 15. See pre-skill 5 in Chapter 2 for the preliminary work that leads pupils to understand this key concept. Once understood, a missing-addend formulation can be modelled on an empty number line, using either of the methods of recording described in Activity 4 above. On a number line, the idea of 'equalising', i.e. *Find what must be added to X to make Y*, can be paraphrased as 'difference', i.e. *Find the gap between X and Y*.

For example, the problem 15 − 7 can be read as *15 minus 7* or *Subtract 7 from 15*, and then paraphrased as *7 and what equals 15* or *Find the gap between 7 and 15*. For the method shown below at the left, the procedure is: mark the number 7 onto the number line and label it below the line, draw one jump to bridge through 10, record the size of the jump (i.e. the complement fact), draw a second jump to reach the target number, label the end of the second jump under the number line, calculate and record the size of the second jump, finally find the answer to 15 − 7 by adding the two jumps.

For the method shown below at the right, the procedure is: mark and label the number 7 and the number 15 on the number line, draw a single large jump between the two numbers, subdivide the single jump into two jumps that bridge through 10, calculate and record the size of each jump, finally find the answer to 15 − 7 by adding the jumps. This method has the advantage of transferring the whole question to the number line before embarking on finding the solution, but requires pupils to have enough spatial awareness to set it out on the page so as to leave a suitable amount of space for the subsequent workings.

Example: 15 − 7

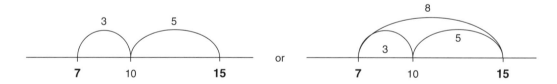

Notice that once a problem has been solved correctly on a number line it is no longer possible to tell from the diagram alone whether the original question was expressed as an addition or as a subtraction. You can see this clearly by comparing the examples in the previous two illustrations. It is one of the great strengths of the empty number line model that a single key method is all that a pupil needs to learn to use as a universal calculation strategy, and that practising the technique reinforces the mathematical connection between addition and subtraction. Make sure to point out explicitly to pupils that the original question dictates where on the number line diagram the answer lies: the answer to an addition problem lies in the furthest number reached on the number line; the answer to a subtraction problem, or a missing-addend problem, lies in the sum total of the jumps.

Activity 6

Practise visualising a number line for mental calculations.

After solving a few problems from Activities 4 and 5 above, pupils should cover up the paper on which they recorded their number line workings and solve the same problems again by visualising a number line in their mind's eye.

Encourage pupils to visualise all the steps of the sum, but only articulate the running total. So, for example, for 8 + 6, the pupil visualises the number 6 as being split into 2 and 4, but says neither of those numbers aloud. Instead, coach the pupil to say only: 8 ... 10 ... 14.

Three ways of visualising the number 6 partitioned into 2 and 4.

Different pupils may prefer different ways of visualising a number being partitioned into component chunks. Some may need to evoke dice patterns or rods; others may have internalised the component facts so thoroughly that they can manage with digits. For more on visualisation, see Activity 13 in Chapter 4. Note that although pupils can choose to visualise whichever partitioning method they find most congenial, the actual addition of the two components is probably best achieved on a virtual number line conjured up in the mind's eye. Working on an empty number line is such a powerful mental calculation tool that we need to build up a virtuous cycle in which pupils use empty number lines at every opportunity so as to become thoroughly familiar and confident in their use.

Do not be tempted to skip this step of the teaching process. It is important for pupils not to become dependent on either concrete apparatus or pictorial constructs. The purpose of apparatus and diagrams is not to dictate a recipe for reaching correct answers, but to provide learners with a lucid, flexible and easily visualised representation that can serve as a cognitive model.

What to teach next?

See Chapter 4 for step-by-step activities to teach pupils to bridge through numbers greater than 10.

CHAPTER FOUR

Bridging through multiples of 10

Overview

As already discussed in the overviews to the previous two chapters, bridging is the powerful mental calculation strategy by which addition is performed as a linear movement on a number line. Crucially, the movement along the line is always in the forward direction, and is never performed one step at a time but instead in component jumps, with the number 10 or a multiple of 10 acting as the stepping stone between jumps.

As I have mentioned previously, although bridging is an abstract concept, I believe that pupils should explore the idea concretely with Cuisenaire rods before learning how to record the work using paper and pencil on an empty number line. Recording is simplified for pupils who have been taught always to work forwards, from left to right, as there will be no need for them to include a plus or minus sign before a number (except when working with negative numbers) nor to attach arrows to the jumps.

For pupils who have a strong number sense, bridging through multiples of 10 can be easily learned as an extension of the basic bridging technique. However, dyscalculic pupils and pupils with other specific learning difficulties do not readily see the repeating patterns within the number system and need to be introduced gradually and systematically to the application of this increasingly familiar technique to larger numbers.

The process of bridging through multiples of 10 can require more than just two jumps. Encourage pupils always to aim for the most effective and transparent methods of calculation, which normally means minimising the number of calculation steps.

Remember that the bridging technique need not be confined to addition problems. Bridging is particularly useful as a way of subtracting by means of complementary addition. Just because addition can be taught as a movement from left to right along a number line, it does not follow that subtraction should be taught as a movement in the opposite direction. Pupils with specific maths difficulties find it painfully difficult to work backwards, and should never be required to do so except, if they are comfortable with the idea, for only one or two small steps of calculation. A backward movement of more than two steps puts too great a burden on the working memory of most pupils with specific learning difficulties such as dyslexia, dyspraxia or dyscalculia.

Before learning the bridging technique, pupils must first have mastered certain pre-skills. See Chapter 2 for details of the pre-skills and for suggestions on teaching them to older learners who have not yet mastered these essential concepts.

This chapter provides a step-by-step guide for teaching bridging through multiples of 10 to learners who struggle with maths, through a series of consecutive activities. The numbering of the activities begins at 7 because they follow straight on from the 6 activities detailed in Chapter 3.

Summary of activities for learning to bridge through multiples of 10

Activity 7. Add a 1-digit number to a 2-digit number. Group together a batch of problems in which the only difference is the value of the tens.

Activity 8. Add a 1-digit number to a 2-digit number. Group together a batch of problems in which the same 1-digit number is being added.

Activity 9. Add a 1-digit number to a 2-digit number. Group together a batch of problems in which only the value of the units in the 2-digit number remains unchanged.

Activity 10. Discuss the commutative property of addition again, in relation to 2-digit addends.

Activity 11. Add a 1-digit number to a 2-digit number on a number line.

Activity 12. Use number lines for solving 2-digit subtractions by complementary addition, keeping the solutions below 10.

Activity 13. Mentally add a 1-digit number to a 2-digit number.

Activity 14. Use visualisation techniques to solve simple subtractions mentally.

Activity 15. Add a 1-digit number to a 3-digit or 4-digit number, initially on a number line and then mentally.

Activity 16. Add two 2-digit numbers for which only the numbers in the units position require bridging.

Activity 17. Use number lines for solving subtractions in which all the numbers, including the solutions, are numbers above 10.

Activity 18. Add two 2-digit or 3-digit numbers for which the necessary bridging is in the tens' position, through 100 or through a multiple of 100.

Activity 19. Use complementary addition and successive bridging to subtract a 2-digit number from a 3-digit number.

Activity 20. Mixed practice.

Activity 21. Bridge through zero when adding onto negative numbers.

Bridging through multiples of 10

Before Activity 7

Ensure all the pre-skills are in place (see Chapter 2) and that pupils are secure with the technique for bridging through 10 (see the six activities in the previous chapter, Chapter 3).

Activity 7

Add a 1-digit number to a 2-digit number. Group together a batch of problems in which the only difference is the value of the tens.

Start by reminding pupils of a sum from Activity 1 or Activity 2, for example 8 + 5, modelled with rods. Immediately after solving that problem with rods, change the problem to 18 + 5.

8 + 5

The solution to **8 + 5** is achieved by bridging and partitioning the 5 into 2 + 3.

18 + 5

The solution to **18 + 5** is also achieved by bridging and partitioning the 5 into 2 + 3.

Next, arrange the same rods in a tens and units formation, so that the rods are positioned vertically with the orange tens rods to the left of the units rod. This formation makes it really easy to see that the second addend must be partitioned into *the complement and what's left*.

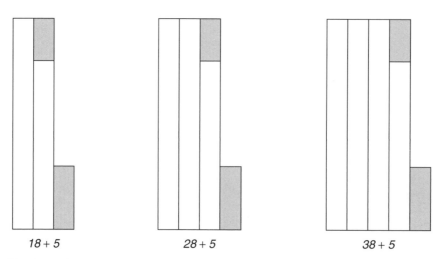

18 + 5   28 + 5   38 + 5

Numbers above 10 are best arranged in a tens and units formation for bridging.
Connect the problem **8 + 5** to **18 + 5** and then to **28 + 5, 38 + 5**, etc.

(Continued)

(Continued)

Have pupils solve related problems such as 28 + 5, 48 + 5, 38 + 5, etc.

Give pupils practice based on another problem encountered in Activity 1 or Activity 2, for example 8 + 7. Model the sum with rods arranged in a tens and units formation. Extend the same set-up to model 18 + 7 and then other related problems such as 58 + 7, 28 + 7, etc.

Continue in this way, giving batches of practice in adding a 1-digit number to a 2-digit number, changing only the number of tens from question to question within each batch.

Activity 8

Add a 1-digit number to a 2-digit number. Group together a batch of problems in which the same 1-digit number is being added.

Model a problem such as 26 + 7 with rods. Arrange the rods for the 2-digit number in the tens and units formation, with the orange rods in a block to the left of the units rod. Such an arrangement mimics place value columns (since the tens are collected and shown to the left of the units) at the same time as making it clear that the second addend must be partitioned into *the complement and what's left*. Continue with a batch of other problems in which the same second addend is added to a variety of 2-digit quantities, e.g. 19 + 7, 64 + 7, 38 + 7, 157 + 7, etc.

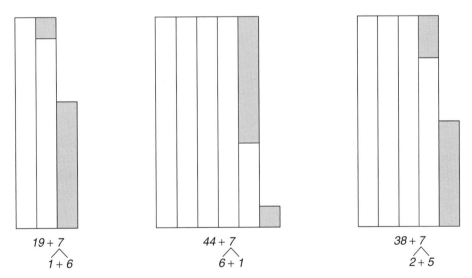

| 19 + 7 | 44 + 7 | 38 + 7 |
| 1 + 6 | 6 + 1 | 2 + 5 |

Although the same amount, 7, is being added to various 2-digit numbers, the 7 must be partitioned in different ways depending on the units-value of the first addend.

Present several other batches of problems designed to give practice in adding the same single-digit number (choosing a number other than 7 next time) to a variety of 2-digit numbers.

This activity reinforces the work in Activities 1 and 2, in which pupils learn that the partitioning required by bridging is performed on the second addend, but is dictated by the value of the units in the first addend.

Activity 9

Add a 1-digit number to a 2-digit number. Group together a batch of problems in which only the value of the units belonging to the 2-digit number remains unchanged.

Continue to use rods. An example of a batch of problems at this stage is: 17 + 6, 27 + 5, 57 + 8, 47 + 7, 77 + 4, etc.

Activity 10

Discuss the commutative property of addition again, now that the pupils are engaged in adding 1-digit numbers to 2-digit numbers.

Use rods to compare at least one pair of additions. Will the answer to 5 + 28 be the same as 28 + 5? [Yes.] Will starting with the 5 be just as straightforward as starting with the 28? [No.]

Remind pupils of their work during Activity 3, in which they discovered that it does not matter which addend is taken first when both addends are in single digits. The advice that all pupils encounter, that it always pays to start addition with the bigger number, only holds true when one of the addends has more digits than the other.

Activity 11

Add a 1-digit number to a 2-digit number on a number line.

Mix up problems from Activities 7–10 and present them on paper to be solved on an empty number line. Unlike the problems met in Activity 4, it is now not 10, but a larger multiple of 10, that serves as the convenient stepping stone between two jumps drawn on the number line. As mentioned in the previous chapter, pupils should be given a choice as to whether they wish to work in two consecutive jumps on the number line, or whether they wish to transfer the whole problem onto the number line immediately and then subdivide the jump representing the second addend into two.

Example: 38 + 7

Activity 12

Use number lines for solving 2-digit subtractions by complementary addition, keeping the solutions below 10.

Remind pupils of their work during Activity 5, in which they learned to see that a subtraction such as 15 − 7 can be solved by finding 7 + ☐ = 15 or by finding the gap between the numbers 7 and 15. Now reframe some of the same sorts of problems as those in Activity 11 above as subtraction to be solved by complementary addition on a number line, making sure that the solutions are all single-digit numbers. For example, the problem 45 − 38 can be reframed as 38 + ☐ = 45, or as *find the gap between 38 and 45*, and then solved on a number line. The workings are identical to those shown in the figure above illustrating the addition 38 + 7. You may need to remind pupils that the solution to the subtraction problem is found by combining the jumps because they add up to the gap between the two numbers, compared to an addition problem for which the solution is found on the number line itself.

Activity 13

Mentally add a 1-digit number to a 2-digit number, using visualisation techniques.

Using the same problems as those in Activity 11 above, challenge pupils to find the solutions mentally. Coach pupils to visualise the problem in their mind's eye while they articulate aloud the partial solutions, i.e. the running total, as they work towards the complete solution. For example, for 38 + 5, pupils should be taught to visualise the 5 being split into 2 and 3, but to say aloud only the numbers *38 ... 40 ... 43*.

The best way to prompt pupils to visualise the problem in their mind's eye is to instruct them to put an empty number line onto a whiteboard in their mind and then work on the line in just the same way as in Activity 11 above. Pupils who find this too difficult need more practice with paper-and-pencil empty number lines.

There is more than one way to visualise the second addend being partitioned ready for bridging. One way, for pupils who are ready to relinquish visual images of dot patterns or rods, is to imagine the digit in question being split horizontally or diagonally into a pair of number components. Pupils who have had a lot of practice with the informal triangular jottings when working on components (see Pre-skill 3 in Chapter 2) may be comfortable continuing with the same format. So for 46 + 7, they would see the sum as it is written here, then imagine the 7 partitioned into the complement of the first number and whatever remains, i.e. into 4 + 3, as shown below at the left, then focus on the three important numbers while they articulate the running total of the calculation: *46 ... 50 ... 53*.

Another successful type of visualisation is to imagine the second addend being partitioned vertically in layers, as illustrated below at the right. Taking the same example of 46 + 7, the pupil can imagine the 7 inscribed on an object with some depth, like a box, which then splits into two shallow layers. The layer at the back, or on the bottom, is the complement (in this case 4) that attaches itself to its complement partner to create a round number, while

(Continued)

(Continued)

the upper layer floats free and consists of the rest of the number (which in this case is 3). You can model this for pupils using both hands placed one on top of the other to represent the whole addend, allowing the top hand to lift off to represent the two superimposed layers that in turn represent the two components. This is much easier to demonstrate than to explain in writing.

Example: 46 + 7

$$46 + 7 = $$

$$46 + \underset{4 + 3}{\overset{7}{\wedge}}$$

or

$$46 + \boxed{7} = 46 + \boxed{4}\ \boxed{3}$$

Pupils imagine partitioning the second addend, using their own choice of visualisation techniques, and then articulate only the running total: 46 . . . 50. . . **53**.

It is important to ensure that the questions are always written down and that pupils are not being asked to memorise the problem at the same time as having to work on reaching the solution.

Activity 14

Use visualisation techniques to solve subtractions mentally. At this stage, the solutions should still be in single digits.

Exactly the same visualisation techniques as in Activity 13 above can be practised on subtractions that are solved by complementary addition. Provide plenty of examples for your pupils to try. It is important for pupils to spend at least as much time practising subtractions as additions, since most pupils find subtraction more difficult.

Activity 15

Add a 1-digit number to a 3-digit or 4-digit number, initially on a number line and then mentally.

By keeping the second addend below 10, this activity demonstrates to pupils that the same process is involved when the bridging technique is applied to numbers in the hundreds and thousands. For example, because pupils now know how to add 7 to 8 and to 18, 28, 38, etc., they can also easily add 7 to numbers such as 108, 318, 1638 or 5078. Draw pupils' attention to the fact that, despite the large numbers, all these problems can be solved in only two steps with only two jumps on an empty number line.

In every batch of questions in which the units' digits are kept the same for both addends, the two jumps on the number line will be identical, too. For example, in the examples in the previous paragraph, the two jumps drawn on the number line will in every case be 2 and 5.

(Continued)

(Continued)

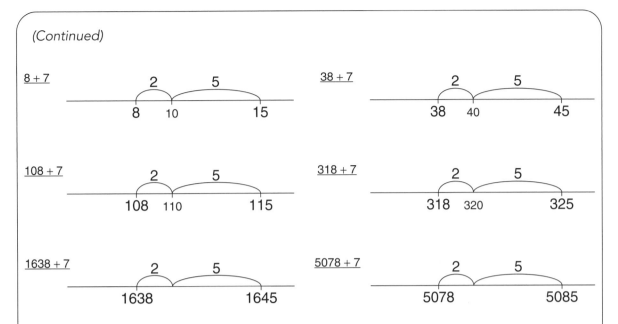

The same sorts of problems can be rewritten as subtraction questions, for example 45 − 38, 535 − 528, etc. Again, because subtractions are solved by complementary addition, only two jumps on a number line are needed.

After finding answers to two or three questions on an empty number line, remove the paper and pencil and ask pupils to use their chosen visualising technique (see Activity 13 above) to solve the same problems mentally.

Make sure that the question is written down so as to avoid putting an undue strain on pupils' working memory.

Activity 16

Add two 2-digit numbers for which only the numbers in the units position require bridging.

Keep the numbers small at first, so that the problems can be easily modelled with Cuisenaire rods if necessary.

Model some sums with rods and model the same sums on an empty number line. While doing both, have pupils experiment with adding the tens first, or adding the units first. There are advantages and disadvantages to either choice, as I will explain below, so I aim to allow pupils to choose their own preference as long as they always stick to the same procedure every time.

Please note that I am not advocating that both 2-digit numbers be partitioned into tens and units. The first addend can be treated as a whole unit – as both addends have the same number of digits, it does not matter which is taken first – and only the second addend need be split into tens and units, or units and tens, depending on the pupil's preference. It would be helpful to present problems at this stage horizontally rather than vertically, so that pupils are not tempted to partition both numbers and attempt column addition.

(Continued)

(Continued)

When pupils are working on an empty number line, you must insist that they minimise the number of jumps on the line, i.e. the number of calculation steps. It is just as important for pupils to practise adding whole multiples of ten rather than single tens as it is for them to practise chunking units together rather than count in ones. For example, the problem 25 + 38 can be solved in three jumps, whether the tens or the units are added first, as shown below on the left. Pupils should not be allowed the kind of notation shown on the right, but must instead learn to partition the second addend into fewer components.

Example: 25 + 38

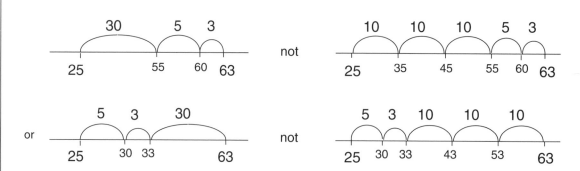

Pupils may add the tens first, or the units first, but must always minimise the number of jumps.

Pupils who cannot easily manage to combine the tens before adding them may need extra practice at this point with the later pre-skills detailed in Chapter 2. The benefits of using a number line are lost when a calculation requires more than four jumps, since the ultimate goal is for pupils to be able to manipulate numbers mentally. The empty number line is not only a paper-and-pencil technique, but is designed to be a useful cognitive model that will lead to efficient, fluent and confident mental calculation strategies.

One great advantage of adding the tens first is that pupils can read both numbers, and the whole sum, from left to right, making it easier to keep track of what to do next and knowing when the solution has been reached. Another advantage is that dealing first with the units of largest value allows pupils to hold on to the actual size of the numbers in question and reach an approximation of the solution after the very first step. This is a general benefit of mental calculation strategies which treats numbers as wholes, over column arithmetic which partitions numbers into place value components and treats each component as units.

However, the advantage of adding the units first is that some pupils can immediately see that the problem can in fact be solved in even fewer than three jumps, as shown below. This can be achieved as a straightforward extension of the kinds of problems met in earlier activities, in which pupils practised adding 1-digit numbers to larger numbers on a number line: *as a first step, add the complement of the first number; as a second step, add what remains of the second number.*

Example: 25 + 38, continued

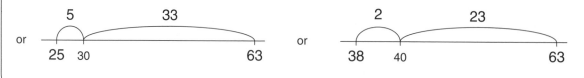

Activity 17

Use number lines for solving subtractions in which all the numbers, including the solutions, are numbers above 10.

Model some subtractions on number lines, taking care to minimise the number of jumps, as discussed in relation to addition in Activity 16 above. All problems presented at this stage should be those that can be solved by complementary addition in only two calculation steps.

For example, 56 – 17 can be solved in two jumps along an empty number line, as shown below at the left. The solution shown at the right is not a sensible alternative, even though the correct answer can be reached by adding the amounts of all five jumps.

Example: 56 – 17

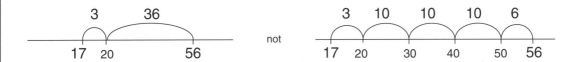

An acceptable compromise would be for a pupil to make three jumps: the first jump for the complement, the second for all the tens, and the third for the remaining units. More concrete work with Cuisenaire rods might be needed on building and partitioning multi-digit numbers (see Pre-skill 10 in Chapter 2) before pupils are comfortable eliding the second and third steps into a single jump.

Pupils need plenty of practice in reframing subtraction questions in terms of complementary addition before attempting to solve them on a number line using paper and pencil. Once pupils can manage these problems in only two steps, they should also try them mentally, using one of the visualisation techniques discussed in Activity 13 above.

Activity 18

Add two 2-digit or 3-digit numbers for which the necessary bridging is in the tens position, through 100 or through a multiple of 100.

Start with problems such as 40 + 80, in which multiples of 10 are added together by bridging through 100. These should be modelled on a number line at this stage, not with concrete materials. Move on to problems such as 90 + 45, or 160 + 73, in which only the first addend is a multiple of 10.

Example: 160 + 73

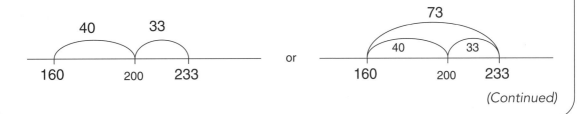

(Continued)

(Continued)

Demonstrate how sums in which the second addend is a round number can be solved in exactly the same way. For example, 86 + 50 is best rewritten as 50 + 86 and bridged through 100 in two steps. This is much quicker and easier than starting with the 86, even though 86 is the bigger number and children will often have been told that it is always easier to start with the bigger number. In this example, it is not at all easy to partition the 50 into 4 and 10 and what's left, because calculating the amount that is 14 less than 50 would require either column subtraction or a separate calculation with bridging on another number line.

Example: 86 + 50

As well as being provided with suitable sums to solve, pupils should be challenged to make up sums that require bridging of the digits in the tens position through 100, or through a multiple of 100. It is often beneficial for pupils to switch their thinking to question formulation, instead of always narrowing their focus to answers for specific questions. Pupils can be given tens and units dice to generate a random 2-digit number, or given a list of various 2-digit numbers, and asked to provide a second 2-digit or 3-digit addend to create a problem that can be solved by bridging the tens through 100. For example, a second addend to follow 67 might be 80, or 250, or any other multiple of ten in which the digit in the tens position is greater than 4.

Strictly speaking, sums in which neither of the addends is a round number, can also be solved by bridging the digits in the tens place, providing the total of the units does not exceed 10. For example, 67 + 82 can be solved sequentially in only two steps and therefore modelled in only two jumps on an empty number line, as shown below.

Example: 67 + 82

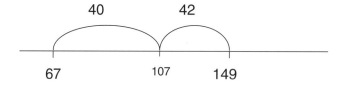

However, the workings are not straightforward because the stepping stone is not a multiple of 10 or 100, and even those pupils who can understand and manage this method usually find it very difficult to identify which problems can be solved in this way. Therefore, I would recommend informal partitioning methods, as shown below, or standard column arithmetic, for solving these types of sums. As Ian Thompson explains[1], partitioning methods lend themselves to being recorded and solved vertically, as column arithmetic, and do not naturally grow out of number line work, an inconvenient truth that current National Strategy guidance fails to acknowledge.

(Continued)

(Continued)

Note that the fact that a problem is being solved using a mainly partitioning approach does not mean that bridging techniques will not contribute to the solution. A problem such as 67 + 82 would require mental bridging for the tens even when set out vertically and worked in columns.

Two other useful styles of recording a partitioning approach are shown at the end of this activity. Both these types of informal notation are suitable models for practising visualisation to support mental calculation. All partitioning methods can be worked either by dealing with the units first (performing column arithmetic from right to left) or by dealing with the digits of greatest value first (from left to right). Pupils should be allowed to choose whichever order they prefer, provided they remain consistent.

Partitioning methods can be usefully modelled with rods first, so that pupils see that all that is involved is a rearrangement of the question, not a new addition (or subtraction) method. For example, 37 + 12 is solved by treating 30 + 10 as one mini-sum, 7 + 2 as another mini-sum, before combining the two subtotals. The actual addition within the two mini-sums may or may not require the bridging technique, depending on the particular numbers involved. Adding 37 + 12 requires no bridging; adding 67 + 82 requires bridging for the tens.

Example: 37 + 12

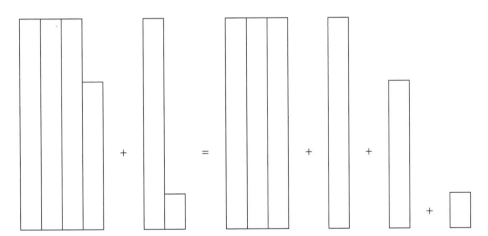

Partitioning methods rearrange the question into separate place value problems.

Example: 67 + 82

Mental bridging is often required when solving problems by partitioning methods.

Activity 19

Use complementary addition and successive bridging to subtract a 2-digit number from a 3-digit number.

Problems to choose for this step are those that require successive bridging, i.e. problems in which bridging is needed both for the units (bridging through 10, or a multiple of 10) and also for the tens (bridging through 100, or a multiple of 100).

Examples of subtractions that require bridging techniques to be used successively are 246 − 78 or 812 − 43. These problems can be solved in only three steps along an empty number line, as shown below.

Example: 246 − 78

821 − 43

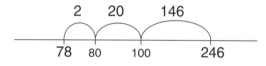

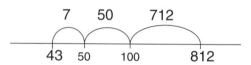

However, many problems that look similar to the ones illustrated above produce jump sizes that are awkward to combine. For example, 275 − 53 'solved' on a number line leaves an addition at the end (175 + 40 + 7) that would require yet more bridging for both the tens and units. By contrast, the same subtraction problem set out and worked in columns is very straightforward because it can be solved by partitioning without any decomposition being required.

In fact, it is worth noting and pointing out explicitly to pupils that the simpler a subtraction problem is to solve by complementary addition along a number line, the more complicated it is to solve by using the standard algorithm in vertical columns, and vice versa.

Example: 275 − 53

$$\begin{array}{r} 275 \\ -\ 53 \\ \hline \\ \hline \end{array}$$

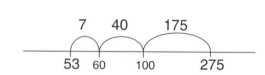

Some multi-digit subtractions are more easily solved in columns than by bridging on a number line.

For those subtraction problems that require decomposition, especially successive decompositions, complementary addition on a number line is far easier than vertical subtraction in columns. The bridging technique on an empty number line is exceptionally helpful when numbers are subtracted from multiples of 100 or 1000, as they often are when working with metric measurements.

(Continued)

(Continued)

Example: 2000 – 53

$$\begin{array}{r} 2\,0\,0\,0 \\ -\ \ 5\,3 \\ \hline \end{array}$$

Bridging on a number line is much easier than a column subtraction requiring decomposition.

See also Chapter 9 for a compensation method for column subtraction from multiples of 100 or 1000, by reasoning that 9 is almost 10.

Activity 20

Mixed practice.

It is one thing to be able to solve problems that a teacher has demonstrated moments before, or to tackle a group of problems that practise a single idea; quite another to be able to choose out of a repertoire of similar techniques the one that is most appropriate for a particular problem. To check how secure your pupils are with the techniques they have learned during the course of their work on bridging, choose problems at random from several different stages of this and the previous chapter, and mix into the selection some addition and subtraction problems that do not require bridging at all.

Activity 21

Bridge through zero when adding onto negative numbers.

The empty number line can be extended to the left beyond zero to show that negative numbers lie on a line that is a mirror image of the positive number line with which pupils are already familiar. Placing negative numbers on the line helps pupils see that negative numbers have a value that is less than zero.

More importantly, using a number line to represent and to visualise negative numbers introduces the idea of direction. Moving from left to right means moving towards ever-larger numbers, i.e. moving in the positive direction; moving from right to left means moving towards ever-smaller numbers, i.e. moving in the negative direction. Thus, a problem such as –5 + 6 does not necessarily need to be rewritten as 6 – 5 but can be solved by starting at –5, which lies 5 steps to the left of zero, and then moving 6 steps in the forward direction. Obviously, pupils should not be allowed to attempt the movement in six separate steps, but should, instead, find the answer in just two jumps, by bridging through zero as shown here.

(Continued)

(Continued)

Example: – 5 + 6

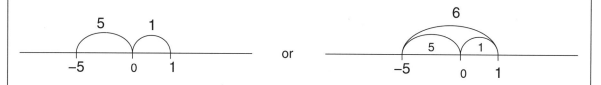

This work paves the way for further work on directed numbers involving adding and subtracting a combination of positive and negative numbers.

Reference

1 I. Thompson (2003) 'Deconstructing the National Numeracy Strategy's Approach to Calculation', ch. 2 in *Enhancing Primary Mathematics Teaching*, ed. I. Thompson, Open University Press.

PART III

The area model of multiplication and division

CHAPTER FIVE

Pre-skills for the area model of multiplication and division

Overview

Multiplication can be equated to repeated addition. Whereas addition can deal with combining different quantities, multiplication deals with replicating a particular number or quantity several times over. This gives rise to repeating patterns that allow the majority of pupils to memorise those multiplication facts that recur most often, i.e. the facts up to 10×10. However, pupils who have weak memories and a tendency not to notice patterns are unable to memorise all the times tables facts reliably, and need to be given the tools to derive the necessary facts for themselves.

Division should be presented as the inverse of multiplication but with connections to repeated addition, not repeated subtraction. Most pupils, even those who are good at maths, find it appreciably easier to work forwards than backwards; pupils with specific maths difficulties are simply unable to count or reason backwards more than one or two steps. Therefore, it makes sense to teach division to all pupils as a process that can be worked in the forward direction, a process by which a number or quantity is repeatedly added until a specified target number is reached.

By the time pupils start secondary school, there are certain pre-skills that should have been thoroughly understood and internalised. This chapter offers some teaching suggestions for those older learners who have not yet mastered these essential skills and concepts.

Summary of necessary pre-skills for the area model of multiplication and division

1. Step counting.

2. Simple mental addition, including sums that require bridging.

3. The ability to see numbers as complete units, not only as collections of ones.

4. Doubling and halving.

5. Place value basics, including decimal numbers.

(Continued)

(Continued)

6. Building and partitioning multi-digit numbers.

7. Times tables facts, or how to derive them from key facts.

8. Understanding what multiplication means and the fundamental connection between multiplication and division.

Note that in the lists of teaching suggestions below, asterisks denote those activities or games that are described in detail in my book *The Dyscalculia Toolkit*[1] and that are therefore not detailed again here.

Necessary pre-skills for the area model of multiplication and division

1. Step counting

Pupils of secondary school age are usually able to step count quite a long way in steps of two, five and ten, without any concrete materials to support the count. For pupils who have trouble step counting in threes and fours, you can use concrete materials to show that step counting is equivalent to repeated addition of equal-sized amounts. In order to count in steps greater than 5, pupils should first know how to bridge mentally, so that they can avoid the trap of counting in ones (see Chapters 2–4).

Suggested teaching activities are:

▶ Practise abstract step counting in 2s, 5s and 10s.*

▶ Cross count, i.e. switch the count from 2s to 10s and to 5s, at random intervals.*

▶ Use concrete materials to support step counting of small numbers other than 2 or 5, for example use Cuisenaire rods in a metre rule in which there is a channel designed to take centimetre-based base-ten materials.

▶ After using concrete materials to support step counting of small numbers such as 3 and 4, pupils practise the step counting abstractly.

▶ Start the count at different points on different occasions.*

▶ Practise two or three forwards steps from a given multiple, but limit a descending count to only one or two backward steps.*

▶ Make an explicit connection between step counting and between creating or reading scales.

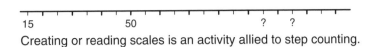

Creating or reading scales is an activity allied to step counting.

2. Simple mental addition, including sums that require bridging

See Chapters 2–4 for a very detailed explanation of how to teach pupils to bridge through 10 and through multiples of 10. Before tackling multiplication and division, pupils must have had enough practice in the bridging technique to be able to add a 1-digit number to any quantity, mentally, through visualising techniques such as imagining an empty number line in their mind's eye.

Suggested teaching activities are:

▶ Learn the complements to 10 by heart.*

▶ Partition numbers below 10 into components, i.e. into chunks, not ones.*

▶ Practise drawing and using empty number lines on paper until the procedure is familiar and automatic.

▶ Games focusing on components (see Chapter 1).

▶ The step-by-step approach to bridging explained in Chapters 2–4.

3. The ability to see numbers as complete units, not only as collections of ones

The best way to achieve the desired shift in focus is to introduce pupils to continuous concrete materials such as Cuisenaire rods. One of the difficulties pupils can have with understanding multiplication is a recurring confusion between 'one' and 'one group'. This may result from a habit of looking at all numbers as if they are simply collections of ones. Pupils should therefore be encouraged to see numbers as complete in their own right. In just the same way as teachers present a ten or a hundred as a complete and distinct entity during work on place value, we can encourage pupils to see all other numbers as a quantity that has been bonded together to form a single cohesive unit.

Some suggested teaching activities are:

▶ Cuisenaire rods activities.*

▶ Notice numbers within numbers, e.g. see that 7 can be built from 4 + 3.*

▶ Build small numbers out of equal-sized groups.*

▶ Play games focusing on components (see Chapter 1).

4. Doubling and halving

Doubling and halving are useful techniques in their own right as well as contributing to handy shortcuts for some multiplication tables facts. It is important to teach doubling as a multiplicative process, an idea that can be confusing to pupils who are in the habit of focusing only on reaching a correct solution, which they may have achieved by means of addition. Using a mirror or other strongly emphasised line of symmetry with concrete material can highlight the kind of thinking we are trying to encourage. Pupils who are not exposed to this sort of visually unambiguous model sometimes believe that *doubling* means the same as *just add another one*. You will know which pupils suffer from this misconception if they consistently give answers that are three times a number when asked to *double and double again* to find four times a number.

When moving on from concrete materials to a diagrammatic model, pupils should continue to emphasise a strong line of symmetry. For example, draw or shade the number of squares representing a particular number onto 1 cm squared paper, then make a thick, maybe coloured, symmetry line before drawing or shading a reflection of the same rectangle.

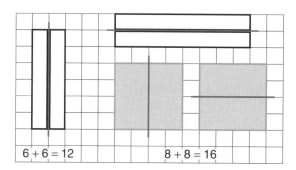

6 + 6 = 12   8 + 8 = 16

Mark a strong symmetry line when exploring doubles on 1 cm squared paper, to emphasise the fact that doubling is a multiplicative process.

Once the doubles up to 10 + 10 are secure, show pupils how to double larger numbers by partitioning them into tens and units before doubling each column value separately and combining the subtotals. Pupils can benefit from an informal notation using arrows, such as the one shown below, rather than a more orthodox layout requiring operation signs. It makes sense for pupils to practise a notation that is easily visualised since the aim is for pupils to be able to perform doubling calculations mentally.

Example: Double 34

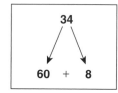

rather than $34 \times 2 = (30 \times 2) + (4 \times 2)$

Example: Double 426

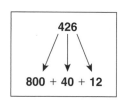

rather than $426 \times 2 = (400 \times 2) + (20 \times 2) + (6 \times 2)$

Arrows allow for an informal notation for doubling that can be easily visualised.

Note that halving is much more difficult than doubling, quite apart from the fraction that results from halving an odd number. The main difficulty is that while a doubling question signals clearly the fact that the process of doubling is required, e.g. 26 + 26 or 26 × 2, the related halving sum can be presented as a subtraction, e.g. 52 − 26, without giving any clue that a doubling/halving fact is involved.

Even when pupils understand that halving is required, they often have difficulty when trying to halve an amount in which there is an odd number of tens. For example, when asked to halve a number such as 36, pupils often proceed by halving the 30 and the 6 separately but then make errors when combining the two subtotals. One way to circumvent this problem is to encourage pupils to work with concrete materials first and partition the number into the teen number and what's left, before halving. Finding half of 20 and 16 separately results in the pair of numbers 10 + 8 that are easier to combine than 15 + 3. Only once the problem is solved concretely should pupils record their thinking with an easily visualised arrow notation, as shown here:

Example: Half of 30

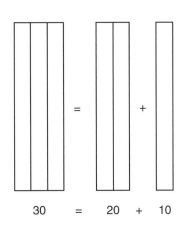

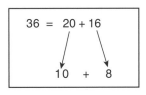

30 = 20 + 10

Example: Half of 36

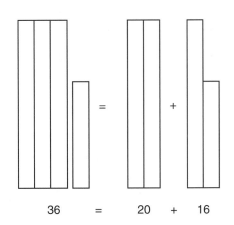

36 = 20 + 16

Some suggested teaching activities are:

▶ Create a doubles pyramid up to 10 + 10 with Cuisenaire rods.*

▶ Use a mirror to emphasise that doubling is based on multiplication rather than on addition.*

▶ Practise halving round numbers, i.e. multiples of ten, paying special attention to amounts in which there are an odd number of tens.*

▶ Practise splitting a 2-digit number in which there is an odd number of tens, into the teen number and what's left.*

▶ Connect doubling to halving.*

▶ Halve a number for which the double has only recently been calculated.*

5. Place value basics, including decimal numbers

As I explain in detail in *The Dyscalculia Toolkit*, and in rather less detail in Chapter 2, many pupils who struggle with maths are confused by the abstract idea of place value. Many have not grasped that place value columns have a logical and threefold repeating pattern of units, tens and hundreds: namely, units, tens and hundreds of *ones*, then units, tens and hundreds of *thousands*, then units, tens and hundreds of *millions*, etc.

To extend an understanding of place value into the columns that represent decimal places, pupils must recognise that the value of a digit in any column is ten times the value of the column to the right, and one tenth of the value of the column to the left. Using concrete base-ten material on place value mats can help foster the notion that every sideways shift of a digit alters its value by a factor of ten. This idea can be especially challenging for pupils with directional difficulties.

Be sure to enunciate clearly, and insist that pupils enunciate clearly, so that everyone can hear the difference between tens and tenths, and between hundreds and hundredths. Use fraction notation rather than lower case letters to label the decimal columns, and encourage pupils to make their decimal point very clearly, perhaps highlighting it with a colour at first.

Thousands			Ones			Decimals		
Hundreds	Tens	Units	Hundreds	Tens	Units	$\frac{1}{10}$	$\frac{1}{100}$	

Note that the threefold repeating pattern for labelling the columns is not as clearly defined in the columns to the right of the decimal point. However, pupils who have been sufficiently exposed to base-ten material will be able to recognise the threefold repeating pattern by focusing on the shapes of the concrete material: cube, long, flat, cube, long, flat, etc.

whole numbers		decimal places		
Ten **10**	One **1**	a tenth **0.1**	a hundredth **0.01**	a thousandth **0.001**

Suggested teaching activities, in addition to those listed under Pre-skills 7 and 8 in Chapter 2:

▶ Connect decimal place value notation to money.*

▶ Build concrete amounts on place value mats.*

▶ Teach ×10 and ÷ 10 as a shift between columns.*

▶ By extension, teach ×100 and ×1000 and ÷ 100 and ÷ 1000, etc., as shifting columns. Rather than teaching pupils about a moving decimal point, I try to get pupils to visualise the place value columns, complete with the decimal point, as an unmovable background, like the mats they have been working on concretely. Digits lie on an invisible layer above this fixed background, and it is the whole upper layer that shifts in the desired direction until the required number of columns have been traversed. This can be easily modelled using transparent overlays.

▶ Locate a decimal number on a number line.* Do not always have the sketch of a line represent the same range, for example the same length of line should sometimes represent the numbers 0–10 and sometimes the numbers between 0 and 50 or 0 and 100 or −50 and 100, etc.

▶ Practise sequencing exercises in which pupils sort numbers with one or two decimal places into ascending or descending order.*

▶ Match various decimal fractions, i.e. numbers including tenths, hundredths or thousandths, to the same number written with a decimal point.

▶ Practise rounding decimals to the nearest whole number, using the same logic as rounding numbers to 10 or to 100.

6. Building and partitioning multi-digit numbers

This pre-skill has already been explored in Chapter 2, where I recommend the use of apparatus such as Cuisenaire rods or base-ten material as a concrete way of investigating the partitioning of large numbers. Multi-digit numbers are commonly split into place value components to aid multiplication, but this is not always the best way. See Activities 11 and 13 in Chapter 6 and Activity 5 in Chapter 7 for more about partitioning.

Some suggested teaching activities are:

▶ Numbers partitioned into tens and units in various ways, using concrete materials before recording the partitions as diagrams or with digits.* For example, 56 split into 50 + 6, 40 + 16, 30 + 26, 20 + 36, and 10 + 46.

▶ Numbers partitioned so that the teen number is split off.* For example, 56 split into 16 + 40, or 88 split into 70 + 18. At first, the numbers should be modelled concretely with Cuisenaire rods.

▶ Numbers recorded as an addition sf tens and units, e.g. 29 = 20 + 9, and also as an addition of multiples of tens and multiples of units, e.g. 29 = (2 × 10) + (9 × 1). At first, the numbers should be modelled concretely.

▶ The Four Throws game.*

▶ Steer the Number game (an Ian Sugarman game[2]).*

▶ The Place Value Boxes game.*

▶ The Calculator Skittles game.*

7. Times tables facts or how to derive them from key facts

Many dyslexic and dyscalculic learners simply cannot remember the multiplication tables facts, no matter how much time and effort they put into trying to memorise them. This is such a common phenomenon that an inability to learn tables is often regarded as a symptom contributing to a diagnosis of dyslexia or dyscalculia. The best way to help such pupils, apart from giving them a tables square from which to read off the answers – a practice I always follow when multiplication or division is not the main focus of the mathematical activity – is to show pupils how to derive an unknown multiplication fact from a tables fact that they know for certain.

As a bare minimum, all pupils must know how to multiply a number by 2, by 10, and by extension by 5. It is perfectly possible to work out any new tables fact from these key facts by step counting up or down. This is what Dorian Yeo[3] calls the 'universal strategy'. The benefit of this approach is that it provides a single method that always works, no matter which table is under discussion.

However, Yeo's 'universal strategy' can be rather cumbersome, entailing as it does so much step counting and a prohibition against 'flipping' tables (i.e. seeing 8 × 4 as if it were 4 × 8). I prefer to teach pupils the area model of multiplication and division from the very beginning, since it, too, is a method that limits the number of facts and procedures that must be learnt by heart. Furthermore, it is an excellent route to encouraging pupils to use logic and reasoning to derive new facts from a small repertoire of known facts (see Chapter 6).

One of the great strengths of the area model is that it highlights the commutative nature of multiplication. This allows the early introduction of certain shortcuts that eliminate the need for unnecessary calculation steps: the idea that every multiple in the 2 times table can be more quickly reached by doubling than by counting up in 2s; the notion that 4 times a number is double the double of the number; the realisation that 5 times a number must be half of 10 times the number; and the recognition that 9 times a number is one step less than 10 times the number. All these shortcuts rely on a thorough exploration, based on concrete experience, of what multiplication means and what times tables represent. These ideas are explored in more detail in Chapter 6.

Some suggested teaching activities are:

▶ Practise step-counting one or two steps up from a given multiple.*

▶ Make tables patterns with Cuisenaire rods in a 10 cm square tray.*

▶ Make tables patterns by shading multiples on a 100-square.*

▶ Learn 9 times a number as one step back from 10 times the number, using knowledge of complements to 10 to minimise calculation.*

▶ Practise using doubling wherever possible as a shortcut, for example double and double again for the 4 times table, and double yet again for the 8 times table, or find 5 times a number by knowing that it is half of 10 times the number.*

▶ Play games with self-correcting cards to learn or practise tables facts.*

▶ Play the Factors games (see CD).

▶ Play the Multiples games (see CD).

8. Understanding what multiplication means and the fundamental connection between multiplication and division

Discover what your pupils understand by multiplication by asking them to use a pile of small items such as counters or nuggets to show what four fives, or 4 times 5, looks like. A pupil who does not really understand what multiplication means, will typically take 9 items and arrange them into two groups, one group of 4 and one group of 5.

Another way of testing for understanding is to pick up, say, three rods of the same colour, perhaps brown, and say, *I have three eights in my hand. But I want four eights. What do I need to add?* Those who lack understanding may suggest you add one more, or offer you a white cube. It is only when they realise that the answer is *one more eight*, that pupils have grasped that multiplication is the name given to the repeated addition of equal-sized groups. This kind of conceptual understanding is more important than the actual numerical answer to a question such as 3×8 or 4×8.

The relationship between multiplication and division should be emphasised from the very beginning. This is best achieved by using both discrete and continuous concrete materials arranged into rectangular arrays, i.e. arranged into a representation of the area model. Pupils must internalise the idea that multiplication and division are nothing more than different points of view when looking at the same relationship. For example, faced with a rectangle made of four yellow rods, it is just as logical, and as easy, to read the rods as *20 can be built out of 4 fives* as it is to read them as *4 fives are 20*.

Rods arranged like this can be read as 20 being built from 4 fives (using the language of division) AND as 4 fives making 20 (using the language of multiplication).

Some suggested teaching activities are:

▶ Build small numbers out of equal-sized groups, arranged as an array.*

▶ Use rectangular arrays of items or rods to highlight the fact that multiplication is commutative.*

▶ Read rectangular blocks or arrays both as multiplication and as division.*

▶ Make simple drawings or sketches to illustrate related multiplication and division word problems.*

▶ Use the inverse of the times tables shortcuts when dividing. For example, if you double and double again to find 4 times a number, think of dividing by 4 as the act of finding half and half again.

▶ Play the Multiples games (see CD).

▶ Play the Factors games (see CD).

References

1 R. Bird (2007) *The Dyscalculia Toolkit*, Sage.
2 I. Sugarman (1997) 'Teaching for Strategies', in *Teaching & Learning Early Number*, ed. I. Thompson, p. 151, Open University Press.
3 D. Yeo (2003) *Dyslexia, Dyspraxia & Mathematics*, pp. 35–263, Whurr.

CHAPTER SIX

The area model of multiplication and division

Overview

Before being able to tackle multiplication and division at the level required for engaging with maths at secondary school, pupils must first have a secure understanding of the concepts and the relationship between the two operations. That is the focus of this chapter. However, before they can deal with the ideas presented in this chapter, pupils must first have mastered at least the first six of the pre-skills identified and discussed in Chapter 5.

It is regrettable that many secondary school textbooks simply presuppose that pupils know their tables facts, and know how to apply them, without ever explaining how pupils might go about acquiring this knowledge. Pupils who have not managed to learn the multiplication tables throughout their years at primary school will need explicit teaching about reasoning methods that do not depend on memorising countless facts and procedures by heart.

This chapter looks at a step-by-step approach to helping learners with specific maths difficulties to understand the concepts of multiplication and division through the area model of multiplication. An important feature of this approach is that multiplication and division are taught side by side, as two aspects of the same construct.

The area model liberates teachers and pupils from the pernicious idea that division means repeated subtraction. In the area model, both multiplication and division can be seen as building up quantities out of replicated groups, with multiplication focusing on the total amount while division focuses on the groups. The only difference between the two operations is where the answer lies: in multiplication, a given number of groups of a given size are built up so that the answer can be found in the total value of all the groups; in division, groups are built up towards a given target number, the dividend, and the answer can be found either in the number of groups or in the size of each group.

The area model of multiplication and division can be applied to any secondary school maths topic that involves either of the operations. For example, it is particularly useful for expanding brackets in algebra (see Chapter 9). It is a model that clarifies the commutative nature of multiplication as well as the all-important relationship between multiplication and division. It is a model that helps pupils see how doubling and halving processes can lead to sensible and serviceable shortcuts. Above all, the area model is a model that is easily visualised, thus providing pupils with a configuration that can be internalised and used as a support for abstract thinking.

A further benefit of using the area model for division is the way that a rectangular array can simultaneously model the two most prevalent ways of thinking about division, namely, the sharing model and the grouping model. It is really important to help pupils understand and distinguish between sharing and grouping (see Activity 12 below) in order to avoid the very common problems that pupils with specific maths difficulties experience with division.

The area model is also ideal for supporting reasoning methods of deriving new facts from known facts.

This chapter contains 15 activities that aim to take the pupils from the concrete stage to the abstract stage at which pupils are ready to understand the maths that lies behind the standard written algorithms for long multiplication and long division.

Summary of activities for multiplication and division

Activity 1. Model multiplication as repeated addition, building up groups of equal size, using discrete items organised into a rectangular array.

Activity 2. Model multiplication as repeated addition, using continuous materials organised into a rectangular array.

Activity 3. Create rectangles by building up rods to match a given total.

Activity 4. Practise reading Cuisenaire rods rectangles both as multiplication facts and as division facts.

Activity 5. Investigate a 10×10 multiplication square, concretely.

Activity 6. Use more transparent language to rephrase a division question.

Activity 7. Make up simple word problems to match Cuisenaire rods rectangles, and practise changing the question from multiplication to division, and vice versa.

Activity 8. Investigate factors and prime numbers, concretely.

Activity 9. Progress gradually to the diagrammatic stage, making sketches of rectangles to represent multiplication and division.

Activity 10. Explore the pattern of multiplication tables on a number line.

Activity 11. Use rectangle sketches to clarify reasoning about tables facts.

Activity 12. Use concrete materials and rectangular arrays and sketches to explain and connect the sharing model and the grouping model of division.

Activity 13. Use rectangle sketches to explore and support the times tables shortcuts.

Activity 14. Make word problems in which the format, the vocabulary and the category are varied.

Activity 15. Explore remainders in division.

The area model of multiplication and division

Before Activity 1

Ensure the necessary pre-skills are in place (see Chapter 5).

Activity 1

Model multiplication as repeated addition, building up groups of equal size, using discrete items organised into a rectangular array.

Take 20 items and arrange them into a rectangular array. Guide your pupils to notice how the same array can be seen both as 4 fives and as 5 fours.

Have pupils repeat this with several other examples from the lower end of various times tables.

Four groups of five.

A rectangular array shows both 4 x 5 and 5 x 4.

Activity 2

Model multiplication as repeated addition, using continuous materials organised into a rectangular array.

Take five purple Cuisenaire rods and four yellow rods. Arrange into rectangles showing 5 fours and 4 fives respectively. Show how, by making a 90° turn, one rectangle fits exactly on top of the other.

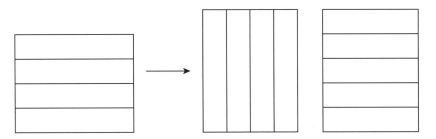

A 4 × 5 rods rectangle is turned to show it can fit exactly on a 5 × 4 rectangle.

(Continued)

(Continued)

Have pupils repeat this activity with several other multiples, for example 4 threes and 3 fours, 3 sixes and 6 threes, 7 eights and 8 sevens.

Note that at this stage you will not be asking your pupils to find the products.

Activity 3

Create rectangles by building up rods to match a given total.

In this activity, which should follow straight on from the previous one, ask pupils to build rectangles out of Cuisenaire rods to match a certain total amount. This is, in effect, a division question, but without the word 'division' being mentioned. For example, ask for a rectangle worth 20 to be built out of fours, or a rectangle made of sixes to be built up to a total of 30. The crucial point that this exercise demonstrates is that division, just like multiplication, can be performed by building *up* groups of the same number. Repeated addition is a much easier concept to understand and to perform than repeated subtraction.

Activity 4

Practise reading Cuisenaire rods rectangles both as multiplication facts and as division facts.

This activity is based on Professor Sharma's teachings.[1]

Introduce the convention of 'reading' rectangular arrays by first reading the vertical side of the rectangle and then the horizontal side. This convention, which follows the practice of Professor Sharma, arises from the fact that the most natural arrangement of rod into rectangles is to stack them horizontally, one above the other, which leads us to use the colour of the rectangle to identify the relevant multiplication table. So, a rectangle built of 5 dark-green rods will be read as five sixes, or five times six, and written as 5×6. Train pupils to point to the relevant side of the rectangle as they say each number. The solution to 5×6 must be reached either by step counting in sixes, never in ones, or by working from one of the key tables facts. In this case, 5×6 is a key fact derived from knowing that the answer is half of 10×6, or half of 60, namely, 30. After making an equivalent rectangle out of yellow rods, pupils read it as $6 \times 5 = 30$.

Next, coach pupils to read the same two rectangles as divisions: $30 \div 5 = 6$, and $30 \div 6 = 5$. Again, insist that pupils use their fingers and hands to relate each of the numbers to the relevant part of the rectangle, as they say each number.

Repeat this activity for many other pairs of rectangles, supplying pupils with the actual products where necessary, until pupils are able to mentally superimpose the related commutative fact onto any area without actually having to build the second rectangle. So, for example, if they build a rectangle of 4 black rods, and read it as $4 \times 7 = 28$, they should be able to turn the rectangle 90°, and read it as 7×4, by imagining seven rows of four superimposed on the four columns of seven that they are actually looking at. Returning to the original orientation, pupils read the rectangle as $28 \div 4 = 7$, and then, after a quarter turn, as $28 \div 7 = 4$.

(Continued)

(Continued)

The 90° turn is not strictly necessary for those who understand the commutative nature of multiplication. However, it can be helpful for two reasons: firstly, to keep to an agreed convention, so that no further elucidation is required over and above a simple numerical statement such as 4 × 5 or 6 × 8; and secondly as a way of connecting this concrete model with the written notation (see Chapter 8).

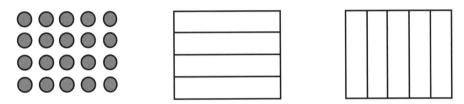

Rectangular arrays, whether built from single items or from Cuisenaire rods, represent both multiplication and division. Each example above can be read as:
4 × 5 = 20, 20 = 4 × 5, 5 × 4 = 20, 20 = 5 × 4, 20 ÷ 5 = 4, 20 ÷ 4 = 5.

Activity 5

Investigate a 10 × 10 multiplication square, concretely.

Many pupils who cannot reliably learn their tables are relieved when allowed to use a multiplication square. I believe it is sensible to give pupils with specific maths difficulties a tables square to use whenever they need it (except when multiplication facts are the main learning objective) but I first want pupils to be able to understand the way this rather abstract piece of apparatus works.

Mark an empty grid on 1 cm squared paper labelling each column and each row from 1 to 10. Have pupils place onto the grid the kinds of rectangles they have already been exploring during the previous activity, for example a rectangle built of 4 fives. The rectangle is placed at the top left of the grid, and an L-shaped piece of card is positioned so as to isolate the rectangle from the rest of the grid, as shown below. Pupils should read the rods rectangle as a multiplication, noticing that the L-shaped card is highlighting the labels on the grid that match the multiplication fact they are reading off the rods, i.e. 4 × 5. Pupils should find the product in the most efficient way they can manage, which may entail counting up four steps of 5, or may be derived by halving 10 × 4. Pupils should next keep the L-shaped card in place and remove the rods. The L-shape can now be seen to be isolating 20 squares from the rest of the grid. The pupil writes 20 in the nearest square to the corner of the L-shaped card, this being the square where the fifth column and the fourth row intersect, and also, of course, the twentieth square of the isolated rectangle, counting from left to right and top to bottom. Pupils should now turn the rods rectangle by 90° and repeat the procedure, ending by writing the product 20 onto the square where the fourth column intersects with the fifth row.

(Continued)

(Continued)

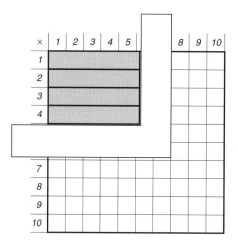

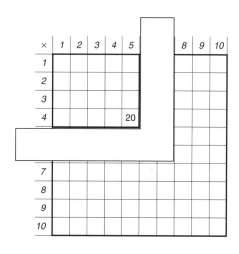

Pupils should not be expected to fill in the whole grid in this way as the grid is too small to accommodate 100 legible numbers. However, they should fill in a few different products, finding both positions on the grid for each product that is not a square number. Finally, have pupils make all the square numbers from 1 to 100, using rods and the L-shaped card as before, and enter all ten products along the diagonal of the multiplication grid.

×	1	2	3	4	5	6	7	8	9	10
1	1			4						
2		4								20
3			9			18			27	
4	4			16	20					
5				20	25				45	
6			18			36				
7										70
8									72	
9			27		45			72		
10		20					70			

Activity 6

Use more transparent language to rephrase a division question presented in digits in writing.

Go back to some of the questions explored in Activities 3 and 4 above, and relate the same examples explicitly to division, using the mathematical vocabulary 'divide' and 'divided by' and the division sign. Teach pupils to rephrase a division question as *How many Xs in Y?* This is a formulation that directly relates to times tables knowledge. For example, when looking at a question such as 32 ÷ 8, pupils should rephrase it as *How many 8s in 32?* They can solve it concretely by building up 8s, using brown Cuisenaire rods, until a total value of 32 is reached, or abstractly by thinking about which step of the 8 times table creates the product 32.

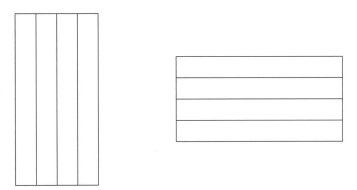

32 ÷ 8 can be understood as having to build up 32 from 8s, i.e. as building a rectangle concretely out of 8s until the value of the whole rectangle is 32. The question can be interpreted as: How many 8s in 32?

If the question were 32 ÷ 4, the pupil can rephrase it as *How many 4s in 32?* in order to relate it to a multiplication tables fact, or, if they already know how to use the relevant shortcut, they can rephrase the question as *Find a quarter of 32*. (See Activity 12 below for more on times tables shortcuts.)

Activity 7

Make up simple word problems to match Cuisenaire rods rectangles, and practise changing the question from multiplication to division, and vice versa.

It is helpful to give pupils practice in making up their own problems before asking them to solve word problems set by others. With a rectangle of Cuisenaire rods in full view, assign a concrete noun to each rod. For example, a purple rod might represent the four wheels of a car, or the four sandwiches in a pack, or the amount of money an item costs, etc. Pupils must use the agreed story set-up to dream up one multiplication and two division problems. For example, looking at a rectangle that is 6 × 4, and after agreeing that the four represents four roses in a bunch, a trio of possible word problems might be: *How many roses would you get if you bought 6 bunches of roses, with 4 flowers in each bunch? If Mum arranges 24 roses into vases with a bunch of 4 flowers in each vase, how many vases will she need? When 24 roses are tied up into 6 bunches of the same size, how many roses are in each bunch?*

This is a surprisingly difficult exercise for many pupils, even when the numbers are kept small and limited to the more familiar times tables. A sample worksheet to practise this exercise can be found on the CD. 💿

Activity 8

Investigate factors and prime numbers, concretely.

Start by using discrete concrete materials, such as counters or nuggets, and ask pupils to arrange them in as many different sizes or shapes of rectangular arrays as possible. For the number 12, for example, 3 different shapes are possible: 12 × 1, 6 × 2 and 3 × 4. Use the word 'factor' to describe the numbers on the sides of the rectangles, e.g. *This rectangle shows us that 3 is a factor of 12 and 4 is a factor of 12. That rectangular array shows that both 2 and 6 are factors of 12.*

Rectangular arrays show that the factors of 12 are 1 and 12, 2 and 6, 3 and 4.

Repeat the exercise with other numbers from familiar multiplication tables, such as 9, 18, 21 or 24. Next, ask pupils to investigate the small prime numbers, such as 5, 7, 11 or 13. They will find that no possible rectangular arrays can be built beyond the straight line that is the number itself. This exercise gives pupils a really clear understanding of what a prime number is.

Prime numbers can only form lines, i.e. an array of 1 times the number itself.

This exercise can be extended by using small cubes to build the rectangles in such a way as to enable the dimensions to be changed by moving rectangular sections, in chunks, to create a new rectangle. For example, build a rectangle from 16 cubes arranged in a straight line. Because 16 is an even number, half of the long thin rectangle can be taken as a chunk of 8 cubes and stacked beneath the remaining half, to create a new rectangle that is half as wide and twice as high as the original. The new rectangle can be treated in exactly the same way, since 8 is also an even number. This kind of rearrangement can be repeated again and again until a long thin rectangle, 1 × 16, emerges, the same dimensions as at the start. The fact that the final rectangle is not in the same orientation as the initial rectangle should make pupils reflect on the commutative property of multiplication.

(Continued)

(Continued)

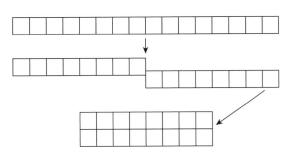

16 units can be shaped into a rectangle **1 × 16.**
Half the rectangle can be stacked to create a new rectangle of **2 × 8.**

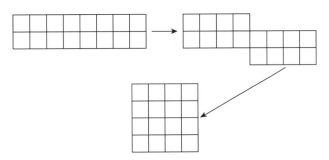

The same process of halving and stacking will create a new rectangle of **4 × 4.**
Repeating the process twice more would result in rectangles of **8 × 2** and **16 × 1.**

Similarly, if one of the dimensions of a rectangle is an odd number, its length might be able to be split into three, five or seven equal parts and stacked to create a new rectangle. For example, a single row of 9 or 15 cubes can be split into three equal chunks. The possibilities for rearrangement end only when one of the sides has the length of a prime number. For example, once a rectangle of 15 cubes has been rearranged from a 1 × 15 into a 3 × 5 format, no further transformations are possible; therefore the manipulation of the cubes has demonstrated that the number 15 has only one pair of factors apart from 1 and 15, namely, 3 and 5.

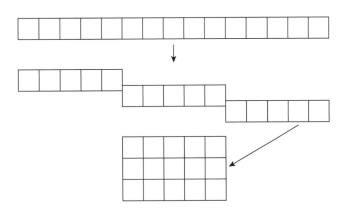

15 units can be shaped into a rectangle **1 × 15.**
This can be split into thirds and stacked to create a new rectangle of **3 × 5.**
No other rectangle is possible; therefore all the factors of 15 are 1, 3, 5 and 15.

(Continued)

(Continued)

To continue this exercise with larger numbers, it is more convenient to use rectangles of paper cut from 1 cm squared paper, than to try to work with too large a quantity of counters or cubes. (You may prefer to postpone this extension of Activity 8 until after Activity 9.) For example, to explore the number 60, pupils can cut out several rectangles that measure 6×10 squares, and try to create other sizes of rectangles by cutting and sticking. Six different rectangles can be created (1×60, 2×30, 3×20, 4×15, 5×12, 6×10). To explore the number 23, a strip measuring 1×23 can be cut out, and pupils will find by trial and error, or by logic and reasoning, that no other shape is possible and that therefore 23 is a prime number.

Activity 9

Progress gradually to the diagrammatic stage, making sketches of rectangles to represent multiplication and division.

Start by using 1 cm squared paper as a base on which pupils build the same Cuisenaire rod rectangles from Activities 3, 4, 5 and 8. Draw round some of the rectangles, then remove the rods and shade the area. Label the sides of the rectangle, and its area. Pupils should practise reading the drawn rectangle, both as multiplication and as division, in just the same way as they previously practised reading Cuisenaire rod rectangles in Activity 4 above.

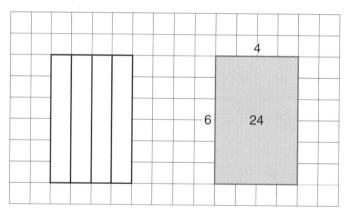

Cuisenaire rods are placed onto 1 cm squared paper at the left.
The same area is shaded on the paper and labelled at the right.

Another way of moving to a less concrete representation of multiplication rectangles is to use small round stickers on 1 cm squared paper with one sticker in each square. For example, ask pupils to build an array of 4 by 5, and to tell you in advance how many stickers they will need, or give pupils 18 stickers and ask them to first estimate, then find, how many 3s in 18 by first imagining and then creating a rectangular array. Stickers contributing to a single rectangle should all be of the same colour.

Finally, but only when they have had enough practice at the earlier stages, pupils will be ready to sketch rectangles, without using squared paper or trying to keep to scale, in order to support their thinking about multiplication and division (see Activities 11–14 below).

Activity 10

Explore the pattern of multiplication tables on a number line.

The process of deriving new tables facts requires pupils to be able to step count mentally from known tables facts. For the 6, 7 and 8 times tables, this often involves mental bridging, which some pupils may find daunting. Pupils who are not entirely secure with mental visualisation of an empty number line can be reassured by first exploring the bridging demands of these three multiplication tables on a structured number line on paper.

A blank skeleton number line for exploring multiplication tables is shown below and on the CD. 💿 The salient features are a starting point of zero, ten jumps of equal size and a line or arrow marking the halfway point of the line.

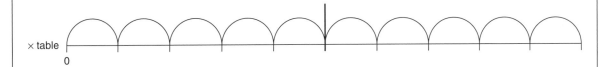

Pupils should start with the 6 times table and label each of the jumps with the number 6. As pupils begin to step count and label the number line itself, it becomes apparent that only four of the ten jumps require bridging, and that there is a symmetrical pattern of jumps on either side of the halfway marker. Furthermore, the number 6 is partitioned in exactly the same way on all four occasions of bridging, namely, into 4 + 2 or 2 + 4.

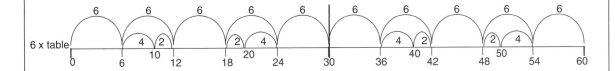

The same exercise should be repeated with the 8 times table before tackling the harder 7 times table. An investigation of the 8 times table shows that there are only two ways of splitting 8, namely 4 + 4 and 2 + 6 (or 6 + 2). An investigation of the 7 times table shows that all three possible ways of partitioning 7 into two components are required at different points of the table, which is precisely what makes the 7 times table so difficult.

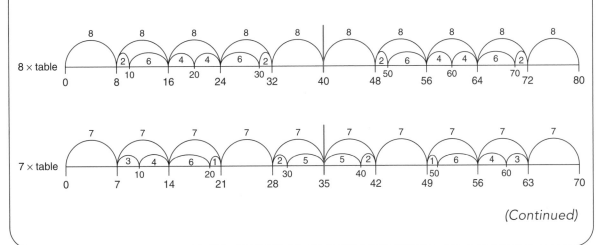

(Continued)

(Continued)

The symmetrical pattern of jumps on either side of the halfway marker will, of course, be present for any multiplication table.

There is nothing to be gained from exploring the 9 times table in the same way, since it is always easier to find 9 times a number by calculating one step less than 10 times the number, a method that requires knowledge of complements but does not involve the bridging strategy.

Activity 11

Use rectangles sketched on paper to clarify reasoning about tables facts.

Note that multiplication tables are first discussed in Chapter 5 as one of the necessary pre-skills for the more general topic of multiplication and division.

Tables, just like other multiplication and division facts, can be represented and visualised using an area model. At first, explore this idea using rods placed on 1 cm squared paper. Draw an outline of, say, the 6 times table onto the squared paper. The drawn rectangle should be 6 squares wide and 10 squares high. Put one dark green rod on the paper at the top of the rectangle to represent 1 × 6. Next, add one rod at a time to show how the table is built up in steps of 6, up to 5 × 6. Revise the answers to these five questions in a random order, by removing and adding rods as required. The higher end of the table should also be practised out of sequence, at first with the required number of rods placed on the paper, and later with the pupil just looking at the 10 × 6 outline drawn on the squared paper, with a thicker line showing the halfway mark, but without any rods actually being present. To answer the question 6 × 6, the pupil starts at the halfway mark representing the fact 5 × 6, and visualises a sixth rod being added to reach the answer of 36. To answer the question 6 × 9, the pupil imagines the rectangle being almost full of rods, so that the answer can be derived by finding one 6 less than 60, which is a complement fact.

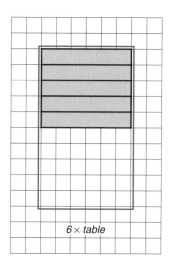

6 × table

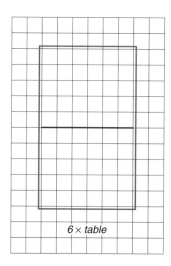

6 × table

(Continued)

(Continued)

An important step is to move towards a more abstract way of working. Make a sketch on plain paper of a rectangle to represent the whole of a multiplication table up to 10 times the number. As the word 'sketch' implies, the drawing should not be on squared paper and need not be to scale. Split the rectangle in two horizontally, remind pupils that 5 times the number is always half of 10 times the number, and note the area of the top half of the rectangle. The sketched outline of the table can now be used to help pupils to work one or two steps up or down from any of the key facts.

For example, if the table being examined is the 7 times table, the outline sketch will show 10×7 and 5×7, from which pupils can be asked to mentally calculate 2×7, 4×7, 6×7 and 9×7. Next, ask pupils to find facts that are more than one step away from the key tables facts, i.e. 3×7, 7×7 and 8×7. Enough practice in step counting gained from working through Activity 10 above should result in pupils being able to answer these questions mentally. Pupils who are still unable to do the necessary mental arithmetic must first return to the addition and subtraction work described in Chapters 2–4.

Repeat the process described above with each individual times table in turn, allowing pupils to sketch only the key facts of a table onto a rectangle to support their mental calculations.

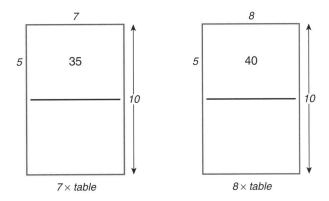

The same sketches of tables outlines can be used for division. For a problem such as $49 \div 7$, discourage pupils from reciting the 7 times table one step at a time until 49 is reached. Instead, show them how to reason from their sketch as follows: *We know that 10×7 is 70 and 5×7 is 35, so will the multiple 49 fall at the lower end (first half) or the higher end (second half) of the table rectangle?* [Answer: higher end (second half) because 49 is more than 35.] *How much more is 49 than 35; is it more than one step of 7 away?* [Answer: 49 is 14 more than 35, that means two steps of 7 more than 5×7. Therefore $7 \times 7 = 49$ so $49 \div 7 = 7$.]

When they are ready to combine questions from different tables, pupils can begin to get in the habit of confining their sketch of a rectangle to represent just the problem they are working on, and not the whole table outline. For example, faced with the question 7×8, pupils may decide that the nearest key fact they know is 5×8. The act of altering a sketch of a rectangle with dimensions 5 by 8 to create new dimensions of 7 by 8, highlights the fact that two 8s must be added to the known fact, and that therefore $7 \times 8 = 40 + 16 = 56$.

(Continued)

(Continued)

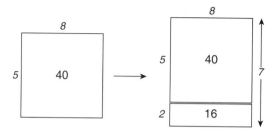

Since pupils should, at this stage, understand that multiplication is commutative, encourage them to label their sketches without thinking about the order or size of the numbers. This may lead to rectangles being altered by extending vertically, as in the example above, or by extending horizontally, as in the example below.

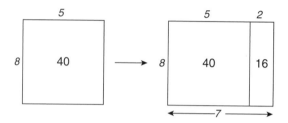

A collection of questions on which to practise this technique of labelling then altering a rectangle sketch might start with examples such as:

$5 \times 7 =$ ___ , so $6 \times 7 =$ ___

$5 \times 8 =$ ___ , so $4 \times 8 =$ ___

and progress to questions just beyond the known times tables, such as:

$10 \times 15 =$ ___ , so $11 \times 15 =$ ___

$5 \times 12 =$ ___ , so $6 \times 12 =$ ___

Pupils can be told that no one expects to know all the tables facts beyond 10×10. Therefore, they can take pride in the fact that the method they are practising for deriving unknown tables facts from key tables facts is exactly the same method that mathematicians with good memories routinely use for deriving multiplication facts that they do not know by heart.

Finally, present pupils with some written mixed multiplication and division problems, all within the known times tables. For example, 3×9, $24 \div 6$, 5×6, $15 \div 5$, $25 \div 5$, etc. Challenge pupils to solve these without using either concrete materials or paper and pencil, perhaps by just imagining Cuisenaire rods being built up into rectangles, or by visualising sketched rectangles of the kind they have been practising.

Activity 12

Use concrete materials and rectangular arrays and sketches to explain and connect the sharing model and the grouping model of division.

There are two models of division that are most commonly taught at school: one is a sharing model and the other is a grouping model. The sharing model is the one that pupils meet first, often during their pre-school years, and involves distributing an amount fairly, usually one at a time. The grouping model is the method taught at primary school as the inverse of the multiplication facts learned through multiplication tables. Later still, pupils are taught that fractions are allied to division but they are not usually informed that finding fractional amounts is related to the sharing rather than the grouping model of division.

To clarify the difference, let us take the example 24 ÷ 3. A sharing approach would have the pupil interpret the question as *Share 24 between 3*, and might see the pupil physically take 24 items and dole them out, one at a time, creating three portions. The answer is found in the value of each equal share. The unit of measurement of the answer is the same as for the dividend, meaning that if it is, say, sweets that are being shared, the answer to a sharing problem will be a certain number of sweets. By contrast, the grouping approach would have the pupils interpret the question as *How many 3s (groups of 3) are in 24?* A pupil who does not know the 3 times table by heart might have to take 24 items and arrange them into groups of 3 until the 24 are exhausted. The answer is found in the number of equal-sized groups.

It is common for mathematically insecure pupils to confuse the two approaches to division and it doesn't help them when teachers unthinkingly switch from one to the other without signalling the fact, and often without even noticing. For example, a pupil who is trying to find half of 18 will not be helped by being advised to think where 18 comes in the 2 times table, unless they have previously been made aware of the connection between the two division models. Similarly, a pupil faced with the problem of finding how many teams of four can be created from 52 players, will not understand how the solution can be reached by finding a quarter of 52, unless the connection between the two models of division has been thoroughly understood and internalised.

Pupils must be taught both models of division, since both are needed in different situations, and should be given the opportunity to explore both in a concrete manner. A good way of connecting the two models is to have pupils practise a way of sharing other than by doling out one item at a time: demonstrate that it is quicker and more efficient to work in rounds than to work with individual items. This means setting aside just enough for one item each, ready for distribution at some future time but without doing the actual distribution, and repeating the process until all the items are exhausted. The result is several groups of items, with each group representing one round of sharing; therefore the number of groups also represents the value of each share.

One of the many strengths of the area model of division is that it makes no distinction between grouping and sharing. A rectangular array of 15 discrete items simultaneously shows all these facts:

3 groups of 5.

5 groups of 3.

(Continued)

(Continued)

15 split into 3 equal shares.

15 split into 5 equal shares.

15 split into groups of 3.

15 split into groups of 5.

A sketched diagram of a rectangle measuring 3 by 5 is a visual way of representing and modelling all the same facts.

Activity 13

Use rectangle sketches to explore and support the times tables shortcuts.

Activity 11 above has already introduced rectangle sketches to support pupils' reasoning about derived tables facts. Rectangle sketches can now be used to model the shortcuts based on doubling and halving. For example, 4×7, a fact that pupils already realise can be derived by working back one step from 5×7, can also be derived from doubling and doubling again. The advantage of the doubling shortcut is that no backward movement is necessary and no bridging is involved. The same logic can be used to derive 8×7 through doubling, which is one of the tables facts that is very often answered incorrectly.

Pupils can choose to sketch their rectangles vertically, as shown here for 4×7 and 8×7, or horizontally, as shown here for 4×8 and 8×8. The importance of the sketches lies in the fact that they support the reasoning process that lies behind the shortcuts. As always, the aim is not to give pupils recipes that will produce answers mechanically; the aim is for them to understand the arithmetic logic that lies behind a calculation procedure.

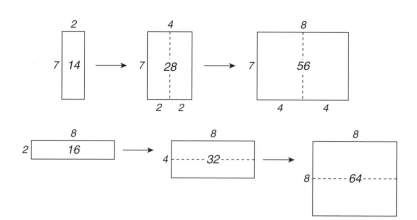

Doubling is a useful shortcut for finding 4 times or 8 times a number.

Sketches can be used alongside the informal arrow notation proposed in Step 4 of the previous chapter.

(Continued)

(Continued)

Rectangle sketches will also support pupils' thinking when reasoning that 5 times a number must be half of 10 times the same number, or when working out 9 times a number from their knowledge of 10 times the same number.

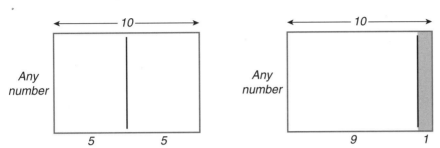

Other useful shortcuts are recognising that 5 times any number must be half of 10 times the same number, and seeing 9 times a number as one 'slice' less than 10 times the same number.

Similarly, rectangle sketches can be helpful for supporting calculations involving a variety of fractional quantities, for example finding a quarter of 48 or a third of 39 or a fifth of 75. When examining the resulting rectangular sketches, discuss with pupils how, when it comes to division, numbers are not always best split into place value components. For example, it is not easy to find a fifth of 75 if the 75 is only seen as 7 tens and 5 units; however, once the 75 is partitioned into 50 and 25, the problem is much more accessible.

Pupils who have learned about fractions through paper-folding exercises[2] may like to draw 'fold lines' on their rectangle sketches at first.

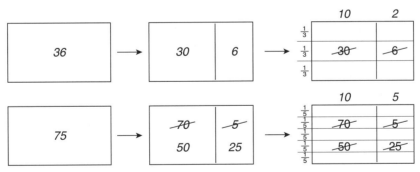

These rectangle sketches model finding fractional amounts: $\frac{1}{3}$ of 36, $\frac{1}{5}$ of 75.

Once pupils understand that finding a third or a fifth entails division, they can proceed as for the more familiar division problems they have practised in earlier activities, in which the answer is found by reasoning from known facts and recorded along the top of a sketched rectangle.

(Continued)

(Continued)

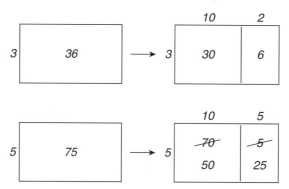

These rectangle sketches model finding fractional amounts: $\frac{1}{3}$ of 36, $\frac{1}{5}$ of 75.

Practice of this kind will lead directly to short division, and later to long division, both of which are examined in more detail in Chapter 8.

Activity 14

Make word problems in which the format, the vocabulary and the category are varied.

It is important to situate multiplication and division problems in the real world. It is also important to vary the vocabulary of word problems so that pupils are not simply responding to 'cue words' like 'groups of' or 'share'. Instead, pupils should develop the habit of visualising the scenario presented in the word problem as a way of understanding how to work towards a solution.

One easy, but underused, way of introducing variety into a word problem is to move the question to the beginning, or the middle, of the word problem instead of always putting the question at the end. For example, compare these three formulations of the same problem. Notice the different positions for the question, and notice the differences in the language that can be engineered for the same simple scenario:

A. *Bread costs £2 per loaf. What will it cost to buy 6 loaves?*

B. *How much money would you need for 6 loaves of bread, if the price of a single loaf is £2?*

C. *You need to buy 6 loaves of bread. How much would you spend at the shop that charges £2 for a loaf?*

The same kinds of variation in language and in the positioning of the question should be sought for division word problems, too.

More fundamental than language and question position is the category of multiplication or division word problem that is being asked. There are at least six distinct types, but only two of them are commonly presented to pupils. The six types can be categorised as:

(Continued)

(Continued)

1. Problems about equivalent groups.
2. Problems about equivalent measures.
3. Rate problems, especially those about the cost of multiple items.
4. Problems about area or rectangular arrays.
5. Scalar problems.
6. Problems involving Cartesian products.

The two most popular categories are problems about equivalent groups, and about the cost of multiple items (as in my example about buying bread, above). However, rate problems other than those about money are far less common. Problems about equivalent measures are closely related to those about equivalent groups, but are rarely posed. Problems about scalar relationships or Cartesian products are very rare indeed.

1. Here is an example of a problem about equivalent groups:

   Pencils are sold in boxes of 6. How many pencils are in 5 boxes? [multiplication]

   Alex bought 30 pencils in 5 boxes. How may pencils to the box? [division].

2. Here is an example of a problem about equivalent measures:

   If four identical parcels are tied up with ribbon, how much ribbon could you afford to use on each parcel out of a 3.2 m strip of ribbon?

   How much ribbon would be needed to tie up 4 parcels if each parcel requires 80 cm of ribbon?

3. Here is an example of a rate problem that is not about the cost of multiple items:

   A car travels at 40 miles per hour. How long would it take to cover 100 miles?

   How many miles could be covered in two and a half hours at 40 miles per hour?

4. Here is an example of a problem about area or rectangular arrays:

   Chairs are arranaged in the school hall in rows of 12. How many chairs will be needed for 15 rows?

   How many rows of 12 chairs can be made with 180 chairs?

Although it is the last example that lends itself particularly to being modelled by the area model of multiplication and division, all the above examples can be approached in the same way, with pupils' thinking being supported by sketches of rectangles.

5. Here is an example of a problem about a scalar relationship:

   Dad is exactly 6 times as old as Sam. If Sam is 8, how old is Dad?
   Dad is 48 and is exactly 6 times as old as Sam. So how old is Sam?

Superimposing two rectangles so that the area of one is represented as being six times greater than the area of the other, can help pupils realise what calculation is required.

6. Here is an example of a problem involving a Cartesian product:

   Stripy T-shirts are made in three sizes and four colours. How many different kinds of stripy T-shirts are made?

   By combining sizes and colours, a factory produces 12 different versions of its stripy T-shirts. The T-shirts come in Small, Medium or Large. How many colours are there in each size?

(Continued)

(Continued)

It is worthwhile modelling such a question, especially the problem expressed as a multiplication question, onto a sketch of boxes within a rectangle, as shown below.

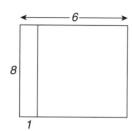

These rectangle sketches model scalar questions and problems of Cartesian products.

Activity 15

Explore remainders in division.

Revert to the very easy tables facts with products below 25 and have pupils model them using discrete items such as counters, nuggets or small cubes. For example, ask pupils to arrange 15 nuggets into an array that is 3 high, to represent the problem $15 \div 3$. The answer can be seen as the width of the array, i.e. 5. Next, give pupils 16 nuggets to model the problem $16 \div 3$. They will build the same shape and size as for the previous problem, but will have one nugget remaining. This should be placed as if it were the start of a whole new column of threes. The array is no longer strictly rectangular, but shows a multiplication or division fact (the complete rectangle) with a reminder of 1. A more sophisticated way to regard the singleton nugget is to think of it not as a remainder of 1, but as 1 out of the 3 that would be needed to complete the next column: 1 out of 3 is written as 1 over 3, or $\frac{1}{3}$ (see also Activity 8 in Chapter 8).

$$15 \div 3 = 5 \qquad 16 \div 3 = 5\tfrac{1}{3}$$

Repeat with other numbers below 25 divided by a number that will leave a small remainder, e.g. $13 \div 4$, $23 \div 5$, $17 \div 4$, $24 \div 7$, $19 \div 6$, $19 \div 4$, etc. These problems can also be modelled with Cuisenaire rods as an alternative to counters or nuggets. The rods encourage pupils to build up towards the target number in steps, rather than in ones, with the added advantage that a remainder stands out because it is not the same colour as the main rectangle.

(Continued)

(Continued)

$$17 \div 4 = 4\frac{1}{4}$$

$$24 \div 7 = 3\frac{3}{7}$$

Once pupils have understood the process, larger numbers can be sketched on diagrams. For example, to solve the problem 37 ÷ 5, pupils should be able to work out that building up groups of five will allow them to reach 35 and that there will be a reminder of 2 left over, or two fifths.

Finally, make up word problems to match the number questions, and ask pupils to think carefully about what a remainder could mean in everyday situations. Some objects, like apples, bars of chocolate or lengths of string, can reasonably be split into fractional parts; others, like people, vehicles or marbles, cannot. Where no split is possible, pupils must interpret the question to find whether the answer must be rounded up or down. For example, take a problem in which an amount is repackaged in different quantities: *Mum buys balloons in packs of 10 so that she can put 3 each into the party bags of the guests attending the birthday party.* If the question is *How many balloon packs must Mum buy to make party bags for 14 guests?* the answer will have to be rounded up from a little more than 4 to 5. If the question is *How many guests could be invited to the party if she buys 4 packs of balloons?* the answer will have to be rounded down from $13\frac{1}{3}$ to 13. See also Chapter 8 for more on remainders.

What to teach next?

See Chapters 7 and 8 for activities to help pupils to understand and manage the formal condensed written algorithms for long multiplication and division. See Chapter 9 for reasoning strategies to do with multiplication and division.

References

1 M. Sharma (1980) 'Multiplication', *Math Notebook*, vol. 1, no. 9; (1988) 'Division', *Math Notebook*, vol. 6, nos 3 and 4.
2 S. Chinn and R. Ashcroft (1998) *Mathematics for Dyslexics: A Teaching Handbook*, ch. 11, Whurr.

CHAPTER SEVEN

Making the transition from the area model to standard written algorithms for short and long multiplication

Overview

Pupils who are weak at arithmetic are best helped by plenty of practical experience with concrete materials. The act of manipulating concrete materials, together with plenty of talk about what is going on, can cause pupils to make connections between different topics in maths, and to develop an insight into how and why certain calculations produce the desired answers. However, the use of concrete materials is not an end in itself and apparatus should never be relied upon to produce a solution mechanically. The superiority of concrete materials over paper-and-pencil methods, as a way of teaching and learning new ideas, lies in the fact that their use leads pupils to construct sound cognitive models.

This chapter sets out a step-by-step approach to teaching and managing that all-important progression from concrete work to abstract thinking, via activities involving diagrammatic models and visualisation techniques that foster powerful cognitive models.

I do not believe in teaching pupils the steps of the long multiplication algorithm as if it were a recipe they must unthinkingly follow. For a start, pupils with weak memories will not be able to recall all the necessary steps in the correct order. But, more importantly, pupils cannot gain insight by passively following instructions; pupils must be taught for understanding so that if their memories fail them they can still work towards the solution by starting from first principles and using logic and reasoning.

There is a debate about whether pupils must hold on to the actual value of all the digits while they are working vertically in columns on paper. I think that is asking rather too much. Pupils must certainly hold on to the value of numbers when they are engaged in mental arithmetic and when they are working in a linear fashion along a number line. But the whole point, and benefit, of written column arithmetic is that it allows number values to be ignored and all digits to be treated as if they were single-digit amounts. Ian Thompson[1] has written compellingly about the way the National Maths Strategies blur the distinction between the two concepts, causing significant difficulties for both teachers and pupils. Teaching the area model of multiplication is a way of avoiding the pitfalls that Thompson identifies. The area model can provide a solid foundation for the more abstract formalised column algorithms because it ensures that pupils are working with the actual value of the digits without the distraction of artificial and confusing mantras about which digits to attend to next or where to 'add an extra zero' or two.

The area model of multiplication, recorded informally as a sketch of one or more rectangles, is also sometimes known as the boxes method. A more stylised and more abstract notation is commonly known as the grid method.

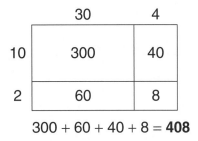

×	30	4	
10	300	40	340
2	60	8	68
			408

$$300 + 60 + 40 + 8 = \textbf{408}$$

The grid method compared to the rectangle (or boxes) area model.

The grid method is an inferior relative of the area model of multiplication. In the most common version of the grid method, pupils are instructed to partition all the numbers into place value components, and to allocate the space of a single line in an exercise book for each place value digit. I dislike the cramped format of the grid and find pupils are often confused about which cells contain numbers belonging to the question and which to the answer, especially as the instructions direct that all the partial products be recorded in yet more cells, producing a dense layout that resembles a spreadsheet full of numbers. I am further troubled by the custom of putting the single-digit number at the top and allocating the 2-digit number to the side, which is a 90° rotation from the standard written algorithm for multiplication, and a complete reversal from the standard written algorithm for long division. Furthermore, when two 2-digit numbers are multiplied, the pupil is sometimes supposed to add partial products horizontally and sometimes vertically, before combing them in a final addition to reach the total product, an unnecessarily complex and confusing procedure that invites errors and hides the logic behind the actual mathematical calculation.

For all these reasons, I favour the more flexible rectangle and boxes method over the grid method. The rectangle area model allows pupils to use diagrams of boxes to support their own reasoning, it encourages them to think carefully about the best way to partition numbers for the most efficient results, and it does not dictate to pupils whether to start the calculation with the tens or the units. An extra benefit is that many pupils find it helpful to sketch larger rectangles to represent larger amounts, even when they fully appreciate that the sketches never have to be drawn to scale. Boxes, rather than the grid, make a clear distinction between the steps of the calculation that depend on multiplication and the final step that depends on addition, and also allow pupils to choose whether to add the partial products mentally, on a number line, or vertically in columns.

By the time they reach secondary school, pupils are expected to be able to solve short and long multiplication problems that are set out in vertical columns. I agree with the government guidance papers stating that pupils are entitled to be taught efficient written methods during their years at primary school. However, I recommend only teaching the abstract vertical column layout to pupils who are already comfortable with the diagrammatic stage and have understood the maths that lies behind it. Furthermore, I do not insist that pupils practise the standard vertical algorithm, no matter what their age, if they themselves prefer to stick with rectangle sketches and the boxes method. For any multiplication problem, whether it arises from everyday life or from an exam, an accurate solution found by sketching areas is a perfectly acceptable and efficient alternative to a solution derived from columns of digits. For those pupils who feel the need to learn the standard vertical layout, I recommend that they be taught a slightly expanded version of the traditional algorithm, as I explain in Activity 3 below.

Summary of activities leading towards written multiplication

Activity 1. Put multiplication tables facts into the wider context of multiplication and division.

Activity 2. Sketch a rectangle to model a short multiplication question. Solve by finding the area of separate boxes before combining the partial products.

Activity 3. Make explicit the connection between a rectangle sketch and a more traditional way of setting out a short multiplication problem in columns.

Activity 4. Minimise the number of boxes when multiplying a 2-digit number by a single-digit number.

Activity 5. Flexible partitioning.

Activity 6. Long multiplication.

Activity 7. Restrict the subdivisions of the rectangle to a maximum of 4 boxes.

Working from the area model to the standard written algorithm for multiplication

Before the first activity

Ensure the necessary pre-skills are in place and that pupils have a sound understanding of the area model of multiplication. See Chapter 5 for details of pre-skills and Chapter 6 for a step-by-step approach to teaching the concepts of multiplication and division through the area model. In particular, pupils should be comfortable with the idea of using sketches of rectangles to model simple multiplication facts before tackling the work in this chapter.

Pupils should also have a good enough understanding of place value to be able to multiply and divide single-digit numbers by ten or a hundred.

Activity 1

Put multiplication tables facts into the wider context of multiplication and division.

Explain to pupils that times tables facts are generally agreed to be 1-digit numbers multiplied together, up to 10 × 10. Extending either the multiplier or the multiplicand to a number beyond the times tables, results in a problem that we call short multiplication. This arbitrary

(Continued)

(Continued)

labelling can be confusing for pupils who are unable to remember their times tables and therefore make no distinction between tables facts and short multiplication. Long multiplication is defined as a problem in which both numbers to be multiplied extend beyond the times tables.

Because any multiplication beyond the tables facts involves numbers over 10, at least one of the numbers has more than one digit; therefore any sketches we make to model the problem must allow for more than one rectangle, or for one rectangle to be subdivided into smaller rectangular areas or boxes.

For example, take any number multiplied by 12. Show pupils how this works with Cuisenaire rods. Building up rows of 12, with one underneath each other, shows very clearly how the total rectangle is actually split into two smaller rectangles of two different coloured rods: orange and red.

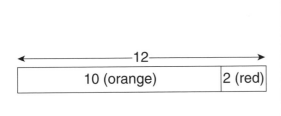

orange	red
orange	red
orange	red
orange	red
orange	red
orange	red
orange	red
orange	red

If 12 is built out of rods, and several 12s are stacked one under the other, the resulting rectangular array is clearly seen as being composed of two distinct rectangles in two colours.

Activity 2

Sketch a rectangle to model a short multiplication question. Solve by finding the area of separate boxes before combining the partial products.

Make a sketch of 7 × 13, as in the example below. Start with one rectangle for the whole 7 × 13, then subdivide it, as shown. At first, pupils might want to mark onto the sketch the seven rows of 13, to mimic the way the concrete rectangle is built out of rods. As soon as possible, encourage pupils to progress to emptier sketches, as shown in the second example at the right.

Example: 7 × 13

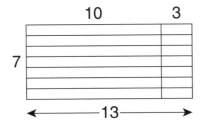 or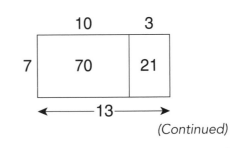

(Continued)

111

(Continued)

Label the areas on the sketch. To find the areas, pupils can be expected to use the multiplication strategies practised in the activities detailed in the previous chapter, namely, their chosen strategy for finding times tables facts together with their knowledge of how to multiply a number by ten or by a multiple of ten.

Ask pupils to explain in their own words why the two areas, the partial products, need to be combined. If necessary, use a highlighter along the perimeter of the original rectangle – the one that represents 7 × 13 – to show that the total area is composed of the two smaller boxed areas. Write 70 + 21 as an addition problem under the sketch, and have pupils find the total mentally. When partial products are not so easy to combine mentally as the example above, pupils should be allowed to write the addition vertically, if they prefer, and to solve it using column arithmetic, as shown below.

<u>Examples: 8 × 15 and 8 × 27</u>

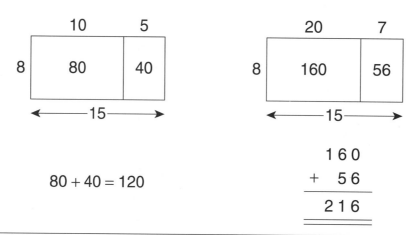

Activity 3

Make explicit the connection between a rectangle sketch and a more traditional way of setting out a short multiplication problem in columns.

At this stage, pupils can be introduced briefly to an expanded version of the standard multiplication algorithm set out vertically in columns. All examples at this stage, i.e. short multiplication, will be composed of 2-digit numbers multiplied by a single-digit number. Make sure to choose numbers in which most of the digits are below 5, so that pupils are able to focus more on the logic of what they are doing rather than dissipating their energy on finding the actual products. Alternatively, provide pupils with a multiplication square (assuming they have already worked through Activity 5 from the previous chapter).

An expanded vertical notation is where the partial products are recorded separately so as to avoid the carrying that may otherwise be required. An example is shown below, and again later in this section. The expanded notation allows pupils to choose whether to start with the digits of greatest or least value, and allows the partial products to be added as column arithmetic, vertically. If a 2-digit number is being multiplied by a 1-digit number, there will be two partial products under the line; a 3-digit number multiplied by a 1-digit number will produce three partial products. The procedure is straightforward and reasonably transparent, especially when it is introduced alongside the area model of multiplication, as suggested here. I prefer this approach to the often recommended procedure of expanding the horizontal written notation

(Continued)

(Continued)

by partitioning the 2-digit multiplicand into place value chunks. The level of understanding that such a procedure would require is exactly what a visual rectangle sketch is so good at illuminating, whereas rewriting a sum such as *24 x 3* so that it reads *(20 x 3) + (4 x 3)* instead, will not strike most pupils as a helpful simplification.

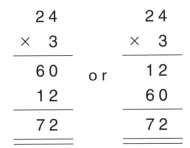
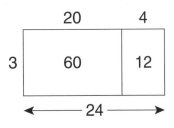

Please note that although it is as easy and as logical to start an expanded short multiplication with the digits of largest value as it is to start with the digits of least value, your pupils' choice may have implications at a later stage. If you intend your pupils eventually to learn the standard condensed written long multiplication algorithm, it will be less confusing if they get into the habit now of working from right to left by starting with the digits of least value. However, because the expanded algorithm shows all the same partial products as the area model, and is such a natural development as pupils progress from the diagrammatic stage to the abstract stage of working, I do not believe that it is necessary to teach the canonical version of the algorithm at all.

At this point, teachers and pupils alike must face the issue about how far it is reasonable to expect to hold on to the actual value of numbers while performing paper-and-pencil calculations in columns. As I mention in the overview at the start of this chapter, I think it is important to be realistic about what each of the different models has to offer. While working through this chapter, which is designed to make explicit connections between the area model sketches and the standard written algorithm, it is important to help pupils make the transition between thinking about actual value and place value. The aim is to lead pupils to a gradual realisation that the standard written algorithm does not require digits to be treated according to their actual values – that is precisely what condenses the algorithm and makes it quick and easy to perform, once it is well understood and well practised – so that pupils can eventually extend the method to larger multi-digit numbers.

For the benefit of pupils who still confuse place value with digit value, it can be a good idea to write Tens and Units headings above the digits of the 2-digit number. The reasoning at this stage for the problem *24 x 3* might go as follows: *3 times 4 is 12, which is written under the line because it is going to be part of the answer; 3 times 2 is 6, so 3 times 2 tens is 6 tens, which is written under the line (on its own line) as 60 with the 6 in the tens place. The answer is the total of the two partial answers, i.e. 12 + 60, which is 72.*

Writing the expanded column multiplication alongside the rectangle sketch highlights the fact that both methods treat the 2-digit number in the same way, i.e. as being built from tens and units, and that therefore both methods generate the same two partial products which are combined to find the total product.

(Continued)

(Continued)

Example: 13 × 7

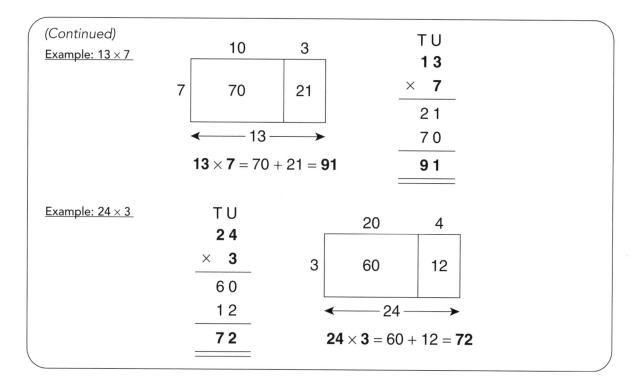

$13 \times 7 = 70 + 21 = \mathbf{91}$

Example: 24 × 3

$24 \times 3 = 60 + 12 = \mathbf{72}$

Activity 4

Minimise the number of boxes when multiplying a 2-digit number by a single-digit number.

When carrying out a multiplication such as 21 × 6, pupils may, at first, choose to split the 21 into three parts: 10 and 10 and 1. Their calculation would look like the example below at the left. However, it can be both daunting and inefficient to have too many partial products to add at the end. Pupils should be encouraged instead to split 21 into only two parts, 20 and 1, and find the solution as shown below at the right.

Example: 6 × 21

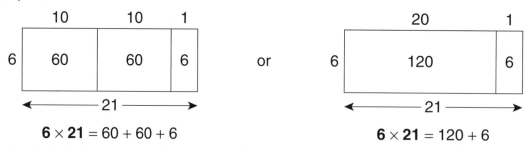

$\mathbf{6} \times \mathbf{21} = 60 + 60 + 6$ or $\mathbf{6} \times \mathbf{21} = 120 + 6$

The logic behind minimising the number of boxes becomes even more obvious as the 2-digit number increases in size. For example, a problem such as 86 × 4 can and should be solved by making a rectangle that is partitioned into two boxes only, resulting in a sum of 320 + 24 to reach the solution. This is clearly preferable to having to find 40 + 40 + 40 + 40 + 40 + 40 + 40 + 40 + 24.

Example: 86 × 4

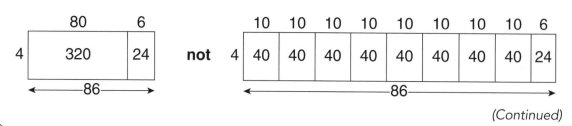

(Continued)

(Continued)

Pupils should continue to practise until they can confidently solve all short multiplication problems in only three steps: finding the area of two boxes within a single rectangle and then adding the partial products to reach the solution.

Activity 5

Flexible partitioning.

The most effective way of partitioning a rectangle into boxes is not always along place value lines. This is one of the advantages that the boxes method has over the grid method, as I mention in the overview to this chapter. For example, a problem such as 28 × 4 can be solved just as easily by partitioning the 4 into two 2s, as by splitting the 28 into tens and units, and leaves a simpler addition at the end. Another example might be 8 × 27, which is more easily solved by splitting the 27 into 25 + 2 than into place value components.

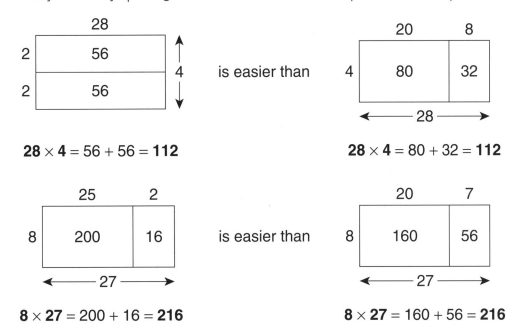

28 × 4 = 56 + 56 = **112**   is easier than   28 × 4 = 80 + 32 = **112**

8 × 27 = 200 + 16 = **216**   is easier than   8 × 27 = 160 + 56 = **216**

The boxes method encourages pupils to be flexible about partitioning.

Similarly, any number multiplied by 5 can be quickly solved by first imagining the rectangle doubled in area, showing 10 times the number, and then halved. This is exactly what we have already taught pupils to do when learning key facts of the multiplication tables.

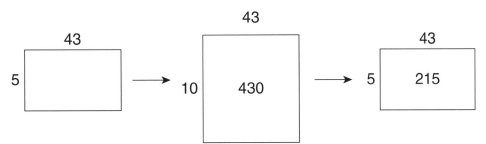

5 times a number can be found by imagining a rectangle being doubled in area to represent 10 times the number, then halving the result.

Activity 6

Long multiplication.

The boxes method should be used with caution when two 2-digit numbers are multiplied together. Sometimes the process is straightforward, as in the first example shown here, while at other times the method can result in a difficult sum at the end, as in the second example. Pupils should be allowed to choose whether to record the addition of the partial products as a horizontal sum to be solved mentally, or as a vertical sum to be solved by column addition. In the second example, column addition is the natural choice.

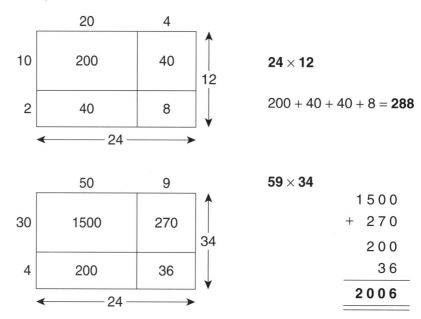

24×12

$200 + 40 + 40 + 8 = 288$

59×34

$$
\begin{array}{r}
1500 \\
+\ \ 270 \\
200 \\
36 \\
\hline
2006
\end{array}
$$

It is at this stage that the traditional multiplication algorithm should be introduced again, starting with the expanded version explained in Activity 3 above, in which each partial product is written on a separate line. Work through the same example side by side, moving back and forth between the boxes and the formal notation at every step. Then, if you feel it is really necessary, you can show pupils how the column method can be further condensed by combining two partial products on a single line. However, there is much to be said for adhering to the expanded long multiplication method and using it for all written column multiplications.

Example: 34×12

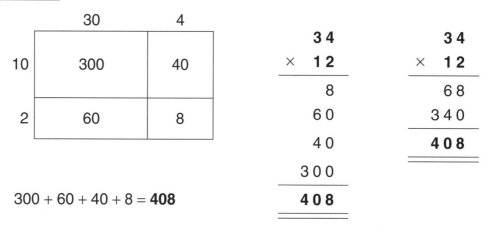

$300 + 60 + 40 + 8 = 408$

$$
\begin{array}{r}
34 \\
\times\ \ 12 \\
\hline
8 \\
60 \\
40 \\
300 \\
\hline
408
\end{array}
$$

$$
\begin{array}{r}
34 \\
\times\ \ 12 \\
\hline
68 \\
340 \\
\hline
408
\end{array}
$$

Performing the same calculation using different methods of notation, side by side, can help pupils understand how the standard written algorithm works.

(Continued)

(Continued)

The box method is excellent for showing pupils what the written algorithms mean and how they work. It can also give confidence to pupils who instinctively prefer to start working on the most significant digits rather than follow the standard method of working from right to left. A very common problem of the standard condensed algorithm occurs because pupils do not understand why it requires them to start with one zero on the second line, with an extra zero on each following line. By contrast, the box method forces pupils to keep hold of the actual value of the products at each stage. The confusion is particularly prevalent when a step of a long multiplication problem results in a multiple of ten, e.g. 62×53.

Example: 62×53

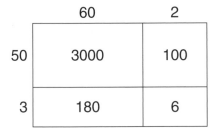

$3000 + 180 + 100 + 6 = \mathbf{3286}$

$$
\begin{array}{r}
\mathbf{6\,2} \\
\times\ \mathbf{5\,3} \\
\hline
1\,8\,6 \\
3\,1\,0 \\
\hline
\mathbf{4\,9\,6} \\
\hline
\end{array}
$$

The area method helps pupils avoid the common mistake shown here at the right.

Activity 7

Restrict the subdivisions of the rectangle to a maximum of 4 boxes.

As mentioned in Activity 6 above, the box method forces pupils to keep hold of the actual value of the digits and the partial products. For precisely this reason, it is not very useful once the numbers become very large. For example, a problem such as 31000×1400 is more easily solved by the traditional column algorithm, correctly performed, than by the rectangular area model. It is even more easily solved with a calculator.

The boxes method becomes rather too cumbersome when multiplying 3-digit numbers with 2-digit numbers or with other 3-digit numbers, because the pupil would be left with the daunting task of adding 6 or 9 partial products at the end. However, the method is still useful for multiplications where a 3-digit number is multiplied by a 1-digit number, e.g. 241×8 or 307×6.

Example: 241×8 Example: 307×6

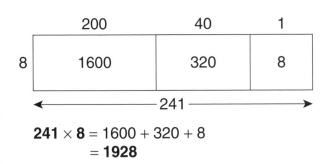   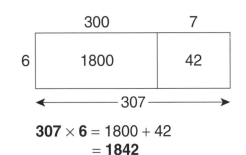

241×8 $= 1600 + 320 + 8$ **307×6** $= 1800 + 42$
     $= \mathbf{1928}$ $= \mathbf{1842}$

(Continued)

(Continued)

As a rule of thumb, if more than four boxes are required to reach a solution, it is time to use the more abstract and condensed standard written algorithms, or to use a calculator. Solving problems using both non-calculator methods side by side will allow pupils to make sense of why and how the condensed written method works, and once that understanding is achieved, pupils can replace the diagrammatic boxes method with the abstract and compact standard written method, when appropriate.

What to teach next?

See Chapter 9 for how to teach pupils to reason about multiplication in order to derive new facts from given facts, and to simplify calculations wherever possible.

Reference

1 I. Thompson (2003) 'Place Value: The English Disease?', ch. 15 in *Enhancing Primary Mathematics Teaching*, ed. I. Thompson, Open University Press.

CHAPTER EIGHT

Making the transition from the area model to standard written algorithms for short and long division

Overview

As I have already made clear in the overview to Chapter 5, I advocate an approach to division that is not based on repeated subtraction, but that is instead closely related to the area model of multiplication. I am aware that my approach runs counter to the advice in the National Numeracy Framework, which currently favours a repeated subtraction model for short and long division, as did the previous National Numeracy Strategy first launched in 1998. I believe that it is better to teach division by working forwards, not backwards, by building multiples up towards the desired dividend in as efficient a manner as possible. Furthermore, I think all pupils should be taught division in this way, not only those with specific maths difficulties.

I hope I have also made it clear by now that I advocate a teaching and learning approach that focuses on teaching for understanding. It follows that I do not believe that any pupils, let alone those with mathematical difficulties or memory weaknesses, should be expected to memorise the standard algorithm for long division and to follow it mechanically through its complex sequential steps. I am surprised and disappointed that prescriptive long division algorithms have made their way back into the maths curriculum in an age when any sensible person would use a calculator instead. However, as long as pupils are going to be faced with difficult paper-and-pencil problems at school, we need to find ways of helping them to succeed. In the case of long division, pupils are best helped by teaching them to understand what division actually means, so that they can solve a division problem by logic and reasoning.

When teaching division, it is important not to conflate the two common models of division by sharing and division by grouping, as explained in Chapter 6. Both models are relevant to long division: a problem such as $1920 \div 48$ is best tackled by finding how many 48s are in 1920 (the grouping model of division), while a problem such as $1920 \div 20$ is best tackled by splitting 192 into two equal parts (the sharing model of division). During the course of working through a long division problem, there are often choices about how numbers are to be partitioned. Therefore, pupils must be familiar with both concepts so that they are able to make sensible choices.

In this chapter, several of the activities contributing to the methodical approach towards formalised written short division algorithms are based on the work of Professor Sharma. I have used Sharma's method successfully with pupils of all ages ever since I saw him demonstrate it some ten years ago. Sharma describes his approach in detail in various articles written for his *Math Notebook* series.[1]

Long division is much more problematic. As Thompson[2] explains, it is very important to link mental processes with written procedures but that the progression from one to the other is not an obvious or natural progression. In Activity 9 below, I explore a way of incorporating ideas about the area model and about long multiplication, in an attempt to make the standard long division algorithm more transparent.

Summary of activities leading towards written division

Activity 1. Revise earlier work on multiplication and division using the area model. Treat division as the inverse of multiplication.

Activity 2. Get pupils to articulate what they understand about the relationship between multiplication and division, in their own words.

Activity 3. Talk pupils through the reasoning steps of a short division problem, through a series of closed questions.

Activity 4. Extend the previous activity to other simple division problems that lie outside the times tables facts, for which no exchange of tens is necessary.

Activity 5. Practise division questions in which tens must be exchanged.

Activity 6. Model the standard notation for short division side by side with a concrete example.

Activity 7. Model the division process using diagrammatic sketches on paper.

Activity 8. How to deal with remainders.

Activity 9. How to manage a long division with an awkward divisor.

Working from the area model to the standard written algorithm for short and long division

Before the first activity

Ensure the necessary pre-skills are in place and that pupils have a sound understanding of the area model of multiplication and how it relates to division. See Chapter 5 for details of the necessary pre-skills and Chapter 6 for a step-by-step guide to teaching the concepts of multiplication and division through the area model.

Activity 1

Revise earlier work on multiplication and division using the area model. Treat division as the inverse of multiplication.

Have pupils build two equivalent rectangles, for example a black and a brown rectangle to show 7×8. Pupils should read the rectangles as multiplication (remembering that the agreed convention is to read the vertical first, as explained in Activity 4 of Chapter 6) and point to where the answer is to be found, which is on the surface of the rectangle. Pupils should next take one of the rectangles they have made and read it both as 7×8 and as 8×7, by simply turning the rectangle one quarter turn after the first reading. In order to find the answer, pupils should not step count in sevens or eights, but instead use the key fact of 5×8 as a starting point, from which 2 further steps of 8 (i.e. 16) must be added.

Using the same rectangles, have pupils read both as divisions. Prompt pupils to show you explicitly where the question is – in the surface area and the vertical side – and where the answer lies – in the horizontal side along the top of the rectangle – by using their fingers to point to the dimensions and the palm of their hand to stroke the surface area.

Activity 2

Get pupils to articulate what they understand about the relationship between multiplication and division, in their own words.

Pupils should by now understand that multiplication is the process by which a number is created from equal-sized groups, and division is the process by which the number is split back into equal-sized groups. We hope that pupils can also see that, just like multiplication, division can be seen as the process of building UP a number from equal-sized groups to the given value. The only difference is that the answer to a multiplication question is the total value of the groups, while the answer to a division question lies in the composition of the groups.

Demonstrate to pupils that rectangles or rectangular arrays are the perfect way to represent division since both division concepts – sharing and grouping – are modelled at the same time (see also Activity 12 in Chapter 6). Pupils should be able to explain these ideas back to you, in their own words, illustrating their explanations with rectangles built of rods and with sketches.

Activity 3

Talk pupils through the reasoning steps of a short division problem, through a series of closed questions.

As an example, the questions you might ask for the problem $36 \div 3$ might go something like this:

(Continued)

(Continued)

▶ *Which is the number you are being asked to divide?* [Answer: 36.]

▶ *So, which number will you start with?* [Answer: 36.]

▶ *What does 36 look like in rods?* Pupils should make 36 out of rods in the most efficient way, i.e. using 3 orange rods and a dark green.

▶ *What do we need to do with this 36?* [Answer: Find how many 3s in the 36, or, Build up in 3s until we reach 36.]

▶ *How can we arrange the 36 to find how many 3s?* [Answer: By making a rectangle that is 3 (units) high.]

▶ *Should we start with the 30 or with the 6? With the biggest rods, or the smallest?* [Answer: With the 30. With the biggest rods.]

▶ *If we take the three orange rods, can we make a rectangle out of them that is 3 high, without having to exchange any of the rods?* Pupils try this out with the rods in front of them. [Answer: Yes.]

▶ *What does the rectangle look like?* [Answer: An orange rectangle, 3 rods (or 3 units) high and 10 (or 10 units, or one ten) wide.]

▶ *How many of the question's 36 are arranged as a rectangle so far?* [Answer: 30.]

▶ *How many are left?* [Answer: 6.]

▶ *What are we going to do with the dark green rod?* [Answer: Turn it into a rectangle that is 3 high, so that it can be attached to (or added to, or included in) the rectangle we've started.]

▶ *Will it fit into the rectangle or do we need to exchange it first?* [Answer: It won't fit as it is. We need to exchange it into smaller rods.]

▶ *Do you need to exchange it into ones?* Once pupils get to this stage, there are three options. Pupils may change the dark green rod into six single units, but they could just as well exchange it into 3 red rods and stack them in the same orientation as the orange rods, or take two light green rods, each one 3 units long, and place them vertically.

▶ *How did you know how to exchange the dark green rod?* [Answer: (if the pupil took ones) I had to exchange the 6 into ones; or (if the pupils took red rods) I needed three rods, to match my three orange rods, and the only way of exchanging 6 into three separate rods is to take red rods, i.e. 3 twos; or (if the pupils took light green rods) I need something to fit into my rectangle that's 3 high, so I needed to exchange the 6 into 3s which are light green, i.e. 2 threes.]

▶ *Have we finished? Is the whole of the question's 36 now in the rectangle?* [Answer: Yes.]

▶ *How many colours has your rectangle of 36?* [Answer: Two colours. Either orange and light green, or orange and red, or orange and white (depending on the pupil's choice when exchanging the 6).]

▶ *Which bit of the question is shown by the orange part of the rectangle?* [Answer: The tens, i.e. the 30.]

(Continued)

(Continued)

▶ *Which bit of the question is shown by the other colour?* [Answer: The units, i.e. the 6.]

▶ *Can you read the dimensions of the rectangle?* [Answer: 3 × 12, or 12 × 3.]

▶ *What number did you start with?* [Answer: 36.]

▶ *Is 36 the area of this rectangle?* [Answer: Yes.]

▶ *What number did the question ask you to divide by?* [Answer: 3.]

▶ *Is 3 the side, or height, of this rectangle?* [Answer: Yes.]

▶ *So, where will you find the answer to the division question 36 ÷ 3?* [Answer: Along the top of the rectangle, or, the width of the rectangle.]

▶ *How wide is the rectangle? Or, How much is along the top of the rectangle?* [Answer: 12.]

▶ *How is the 12 built up?* [Answer: One ten and two units.]

▶ *Can you read your rectangle as a division?* [Answer: 36 ÷ 3 = 12.]

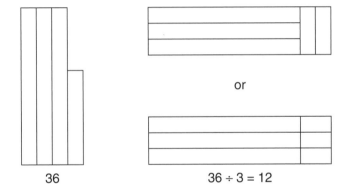

36   or   36 ÷ 3 = 12

This series of questions, based on my personal observation of Professor Sharma's work with pupils, are carefully worded to prompt pupils and steer them in the right direction while ensuring that they are the ones actually doing all the work.

Activity 4

Extend the previous activity to other simple division problems that lie outside the times tables facts, for which no exchange of tens is necessary.

At first, choose problems that lie just outside the times tables facts and do not require any tens to be exchanged, e.g. 48 ÷ 4, 77 ÷ 7, 26 ÷ 2. Later, choose problems with larger numbers that still do not require any tens to be exchanged, e.g. 62 ÷ 2, 69 ÷ 3, 84 ÷ 4, 105 ÷ 5 (build the 105 out of 10 orange rods and one yellow). On each occasion, ask as many carefully directed questions as exemplified in Activity 3 above. The questions serve to coach the pupils into the kind of thinking they will have to do for themselves when tackling both short and long division. This kind of coaching is based on the pupil using familiar logic and reasoning to reach the answer, instead of expecting pupils to memorise and proceed mechanically through the rigid steps of a compact arithmetic algorithm.

Activity 5

Practise division questions in which tens must be exchanged.

Some examples of questions at this stage are: 64 ÷ 4, 51 ÷ 3, 56 ÷ 2, 84 ÷ 7, 72 ÷ 6, 75 ÷ 5.

Use the same questioning technique as shown in Activity 3 above to coach pupils through the reasoning steps towards the solution. For example, here are the questions you might ask for a problem such as 92 ÷ 4:

- *What number is being divided?* [Answer: 92.]

- *What does 92 look like in rods?* Pupils should make 92 out of rods in the most efficient way, i.e. using 9 orange rods and a red rod.

- *What are you going to do with this 92?* [Answer: Find how many 4s in the 92, or, Build up in 4s until I reach 92.]

- *So, you'll be building a rectangle. Do you already know the length of one of the sides?* [Answer: Yes, one side will be 4 (or 4 units).]

- *Will you start with the 90 or with the 2? With the biggest rods, or the smallest?* [Answer: With the 90. With the biggest rods.]

- *If you take the nine orange rods, can you make a rectangle out of them that is 4 high, without having to exchange any of the rods?* Pupils try this out with the rods in front of them. [Answer: Yes.]

- *Have you used all your orange rods in the rectangle?* [Answer: No, there's one orange rod left over.]

- *What does your rectangle look like?* [Answer: The rectangle is orange and measures 4 rods (or 4 units) high and 20 (or 2 tens) wide.]

- *Have you finished? Have you dealt with all of the question's 92?* [Answer: No, not yet.]

- *How many of the question's 92 have you dealt with so far?* [Answer: 80.]

- *How much of the original 92 have you not yet dealt with?* Or, *How much of the question do you still have to arrange into a rectangle?* [Answer: 12.]

- *Where is the 12?* The pupil points to the left-over orange rod and the red rod.

- *The 12 is left over because those rods don't fit into the rectangle. So, what can you do with the 12?* [Answer: It won't fit as it is, so I need to exchange it into smaller rods that will fit.]

- *What are you going to exchange the 12 into?* As before, there are options, but the option of changing the two rods into 12 white unit cubes should be discouraged at this stage, because changing numbers into ones becomes increasingly inefficient as numbers grow larger. Pupils should either choose to exchange the 12 into 4 equal rods and stack them in the same orientation as the orange rods, in which case they will take 4 light green rods, or choose to take as many 4s as will build up to 12, in which case they will take three purple rods and place them vertically.

- *How did you know how to exchange the 12?* The pupil's answer should be in line with their choice, above. If they have chosen 12 white rods, question them now about

(Continued)

(Continued)

whether the 12 can be built in a more efficient manner that nevertheless allows it to be arranged in a rectangle in which the side is 4. Pupils who are unsure about how to improve their choice of 12 single units should be steered towards the grouping rather than the sharing model, i.e. exchanging the 12 into 4s, in preparation for working with numbers that do not divide exactly.

▶ *How many colours has your final rectangle?* [Answer: Two colours. Either orange and light green, or orange and purple.]

▶ *Which bit of the question is shown by the orange part of the rectangle?* [Answer: Most of the tens, i.e. 80 of the original 90.]

▶ *Which bit of the question is shown by the other colour?* [Answer: The remaining 12.]

▶ *Have you finished? Have you dealt with all of the question's 92?* [Answer: Yes.]

▶ *Can you read the dimensions of the whole rectangle?* [Answer: 4×23, or 23×4.]

▶ *What number did you start with?* [Answer: 92.]

▶ *Is 92 the area of this rectangle?* [Answer: Yes, because $80 + 12 = 92$.]

▶ *What number did the question ask you to divide by?* [Answer: 4.]

▶ *Did you make 4 the side of this rectangle?* [Answer: Yes.]

▶ *So, where will you find the answer to this division question?* [Answer: Along the top of the rectangle.]

▶ *How much is along the top of the rectangle?* [Answer: 23.]

▶ *How did you get the answer of 23?* [Answer: Along the top of this rectangle we can see two tens and three units.]

▶ *Can you read your rectangle as a division?* [Answer: $92 \div 4 = 23$.]

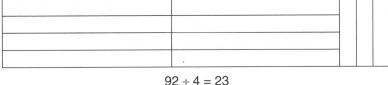

$$92 \div 4 = 23$$

Activity 6

Model the standard notation for short division side by side with a concrete example.

While the pupils go through the same procedure as described above, the teacher demonstrates how each step is recorded using the short division standard notation. First, talk about how a short division problem is typically presented and point out to pupils how the notation can be connected to the concrete work with rods. For example, a problem such as $60 \div 5$ is typically presented as follows when it is to be solved by short division: $5\overline{)60}$. Point out to

(Continued)

(Continued)

pupils how the notation mimics part of the rectangle, showing the top and left sides of the rectangle; how the placement of the number 60 in the notation reminds us that the number to be divided is the area of the rectangle; that the notation tells you that the side of the rectangle must have a height of 5; and that the answer will be the amount along the top of the rectangle.

5 | 60

5 | 60

The short division notation mimics the familiar sketch of the area model of division.

A worked example for the problem 60 ÷ 5 might go something like this:

Step 1. Write $5\overline{)60}$ on the board and point out, as described in the paragraph above, the connection between the notation and the area model of multiplication and division.

Step 2. Ask pupils to take 60 in rods. The most efficient way is to take 6 orange rods. Ask if they can begin to make a rectangle that is 5 high out of the tens they have, without making any exchanges. Pupils can begin the division by making a rectangle of 5 orange rods without exchanging any of the rods. Note that this can only be achieved through the sharing model of division: if the question were interpreted as *How many 5s are in 60?* all the orange rods would have to be exchanged for yellow rods, which would be far less efficient.

Go back to the board. Reminding pupils of the 5-high rectangle they have just built out of five of the six 10s, ask them for the dimension along the top of that rectangle. The answer is 10, or one ten. Write Tens and Units labels in the appropriate place above the written notation, leaving enough room for a written answer beneath the headings, and write 1 in the tens position. Some pupils at this point get muddled between references to 'ten', 'a ten' or 'one ten'. For their benefit, you could, at first, write the whole number 10 on the answer line, marking the zero only very lightly on the board, so that another digit can be superimposed onto the zero at a later stage.

<u>Example: 60 ÷ 5</u>

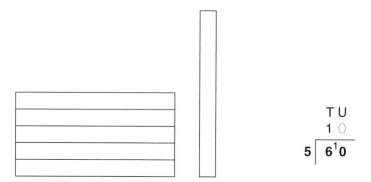

Step 3. Ask how many rods still remain that have not yet been divided, i.e. are not yet part of any rectangle. The answer is that there is a remainder of one ten, i.e. 10. This time, it is clear what the pupil's answer of 'one ten' or 'ten' actually means because the orange rod is physically present. It can be useful to explore this with pupils who muddled place value with

(Continued)

(Continued)

absolute value in step 2 above. Note that it is useful to use the vocabulary 'remain' and 'remainder' here, because there is nothing conceptually different about this remainder, compared with a remainder of units at the very end of a division problem. In response to the pupils' answer of 10, the teacher writes the remainder as a small 1, in front of the zero of the 60, so that the two digits together can be easily read as 10. This is another possible sticking point for some pupils and, again, it is the physical presence of the remaining orange rod that helps pupils understand what is going on: this remaining ten must be moved from the tens column to the units column because there is no possibility of continuing the division without exchanging the ten into smaller rods.

Recap as follows, pointing first to the pupils' arrangement of rods and then to the notation on the board: *We started with 60. You took those 6 orange rods. Here, on the board, is the 60 we are arranging into a rectangle. You made as big a rectangle as you could without making any exchange, and found you could create a rectangle of 5 by 10, with the 5 along the side and the 10 along the top, and with one 10 left over. Here, on the board, is where we record the 10 that is the width at the top of the rectangle, and here is where we record the one 10 that's left over and that we are now treating as ten units.*

Ask the pupils to explain this back to you, referring to the rods and to the written notation in just the same way as you just demonstrated.

Step 4. Ask pupils what they intend to do with the remaining 10 and how it can be fitted alongside the existing rectangle to make a larger rectangle of the same height. Pupils have to make an exchange at this point, but may choose to exchange into 2s, so that there are five separate rods (the sharing model of division), or into 5s, reasoning that the desired rectangle must be five units high (the grouping model of division). Pupils arrange their exchanged rods and extend the orange rectangle into a rectangle that is 5 high and 12 long. The teacher mirrors this activity by completing the written problem on the board, writing a 2 in the units position.

Example: 60 ÷ 5

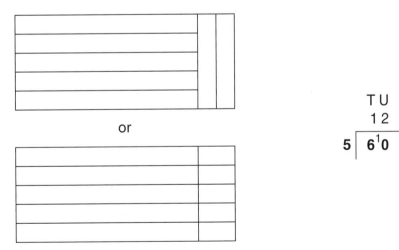

Try a few more examples for which the pupils perform the manipulation of the rods while you record the operation on the board in digits. Next, ask the pupils to direct what you write on the board for a problem that they have not yet solved with the concrete materials. Finally, see if pupils can solve similar abstract problems themselves. Choose problems with relatively small numbers at first, so that pupils who wish to check their answers with the concrete materials will not have too onerous a task.

Activity 7

Model the division process using diagrammatic sketches on paper.

The pictorial stage is a very useful transitional stage between manipulative concrete work and purely abstract work on paper. This idea has already been explored for numbers within the times tables in Chapter 6.

The pictorial stage in relation to short and long division can be modelled for the pupils as follows: *Let's take as an example the problem 78 ÷ 6. The problem can be solved by making 78 into a rectangle with a height of 6, so we will sketch a rectangle and label the side 6. We will start with the biggest values, that means starting with the 7 tens. The largest rectangle we can make out of the whole tens, with the required height of 6, has 10 (or one ten) along the top. We can label the rectangle accordingly. So far, we have taken care of an area of 60, so we must create a second rectangular area for the remaining 18 of the problem. How can we make a rectangle of 18 with 6 on the side? The answer is 3, which we can now use to label the top of the additional rectangle. The dimensions of the whole rectangle can now be seen to be 6 by 13, therefore the answer to 78 ÷ 6 is 13.*

Example: 78 ÷ 6

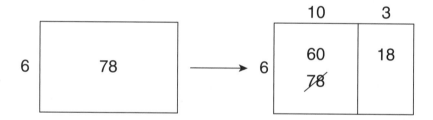

This way of setting the work out on paper has clear connections to the box method of multiplication which pupils have already learned and practised during the work explained previously in Chapter 7.

Pupils should practise using diagrammatic sketches on several problems in which 2-digit numbers are divided by a variety of single-digit numbers, and then practise solving the same problems on paper with both the diagram and the standard notation set out side by side. Give pupils as much practice as they need to help them fully understand the connection between the semi-concrete, i.e. pictorial, and the abstract methods of solving a short division.

Example: 78 ÷ 6

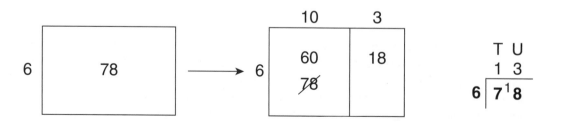

Activity 8

How to deal with remainders.

This activity follows on from the one above and should not be attempted until pupils understand how to manage simple division problems without remainders. Pupils should also be familiar with the term 'remainder' as meaning 'that which remains' after a particular stage of the calculation.

Note that remainders in division are first explored at the end of Chapter 6.

Using concrete materials, start with examples that are similar to the ones already tackled but that will yield a remainder of one or two. After going through the reasoning procedure, pupils will find that they cannot make a neat rectangle from the whole dividend, because there are some rods left over. These rods should be arranged at the right, and be seen as the beginning of a new column. For example, $47 \div 4$, will require 4 orange rods to be arranged into a 4×10 rectangle, and the black rod to be exchanged either for a purple and a light green, or for seven white rods.

Either way, the final complete rectangle measures 4 by 11, but it is immediately obvious that there is a remainder, i.e. an amount that does not fit neatly into the 4-unit high rectangle. In this example, the remainder is 3. But it is also clear from this way of working that the remainder of 3 can be viewed as three out of the four that would be needed for another full column. Show pupils how to write '3 out of 4' as ¾, or three-quarters.

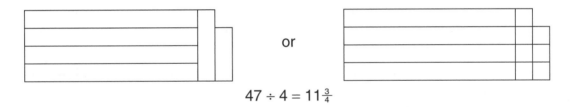

or

$$47 \div 4 = 11\tfrac{3}{4}$$

Tackling the same problem in a purely abstract way would entail setting out the sum as: $4\overline{)47}$. The first question is how 4 tens can be built out of 4s, with the answer of 10 being written as a 1 in the tens column (with or without a faint zero acting as a place holder, as suggested in Activity 6). Since all the tens have now been taken care of, the next question is how to build 7 out of 4s. One complete 4 can be built, producing the answer of 1 in the units columns, but the remaining 3 (or 3 ones, or 3 units) must also be subjected to the divisor, so that the remainder goes through the division operation in exactly the same way as every other component part of the number being divided. The result is 3 divided by 4, which can be written as a fraction by understanding that the line between the top and bottom numbers of a fraction is a division line.

<u>Example: $47 \div 4$</u>

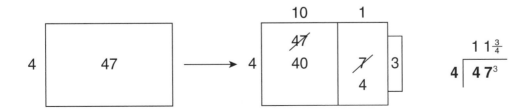

Activity 9

How to manage a long division with an awkward divisor.

So far, the examples presented to pupils have been numbers greater than 10 being divided by numbers less than 10. In true long division, both the dividend and the divisor are numbers greater than 10. This means that pupils cannot simply use multiplication tables facts. Pupils are expected to use their estimating abilities instead, before doing a series of long multiplications alternating with subtractions. This process is a nightmare for pupils with maths difficulties, since, apart from their trouble with subtraction, they very often have extremely weak estimating abilities coupled with memory problems that result in them being unable to memorise the sequential steps of such a complex algorithm.

If I had my way, pupils would be allowed to use calculators for all long division problems. But, if they must learn a paper-and-pencil method, I suggest that it may be possible to get round the problem of remembering the steps of the algorithm by unpicking the process and allowing pupils to discover what each step achieves. This can be accomplished by working through parallel examples in which a single problem is solved on a diagram alongside the formalised written notation, with the pupils moving back and forth from one to the other at every step, in just the same way as pupils have already experienced and practised for short division in earlier activities.

Pupils' other main problem, that of not being able to estimate sensibly, can be avoided by setting out a skeleton multiplication table of the divisor before starting on the division problem itself. This may be time-consuming, but many pupils greet this idea with some relief and feel that the extra time is well spent. All the steps can be calculated in just the same way as pupils have already learned to do for the times table facts, and once the steps are written out, the problem can be tackled as if it were short division.

For example, here is a description of how one might work through the problem 784 ÷ 16 in a way that makes explicit connections between a non-canonical diagrammatic solution based on an understanding of the area model of multiplication/division and the canonical standard written algorithm. The description is very detailed, for the sake of clarity, and is best followed by actually performing each step as you read through the instructions.

Step 1. Create a skeleton table for 16, in an area of the page that is first ruled off to accommodate workings. Against the skeleton, enter the three key facts, 1 × 16, 10 × 16, and 5 × 16 (this last is calculated as half of 10 × 16).

784 ÷ 16	16 × table
	1 × = 16
	2 ×
	3 ×
	4 ×
	5 × = 80
	6 ×
	7 ×
	8 ×
	9 ×
	10 × = 160

(Continued)

(Continued)

Step 2. Sketch a rectangle to represent the division problem and label the height 16. The area of the whole rectangle, which represents 784, is subdivided into two boxes. Provisionally label the left-hand box 700 and the right-hand box 84, leaving space for noting alterations to the figures. Boxes can be added as the calculation progresses at the same time as the areas are re-partitioned and re-labelled.

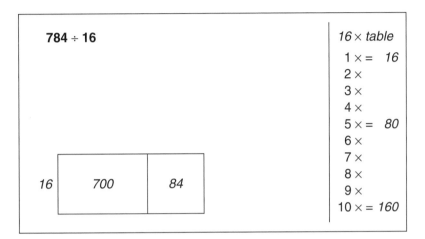

Step 3. Pupils first turn their attention to the digits of largest value, or to the box containing the largest values, i.e. the 700. It is impossible to arrange 7 (hundreds) to create a height of 16, therefore pupils must interpret the 700 as 70 (tens). The table skeleton shows that the entry for 5 × 16 is more than 70, but not as much as 16 more. So, 4 × 16 is as near as one can get to 70. The actual product for 4 × 16 is now calculated, either by working back one step of 16 from 80, or by doubling and doubling the number 16; the answer of 64 should be entered onto the table skeleton in case it will be needed again later. Pupils now need to remember that it is 64 *tens* that has been created by multiplying 16 by 4 *tens*. The area of the left-hand box can now be altered to read 640 and the width at the top of the box is labelled 40. At the same time, the remaining 60 that has not been accounted for in the left-hand box is moved to the next box to the right and added to the 84 already there, creating a new provisional area of 144.

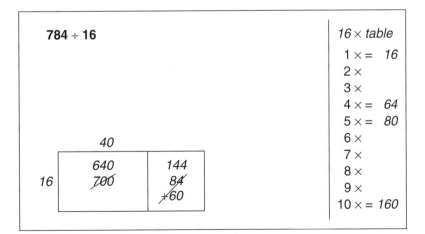

Step 4. At this stage, pupils should set out the written notation alongside the diagrams and label the 3-digits of the dividend with Hundreds, Tens and Units headings, leaving enough

(Continued)

(Continued)

room below the headings to accommodate the answer. By covering up the digits of the dividend one at a time, point out to pupils that the question is first asking them to find how many 16s are in 7, and, failing that, in 78. A look at the skeleton table will show that the closest answer to 78 is the fourth step of the table. Once the number 4 has been written in the tens position, 64 tens must be subtracted from 784 to see what part of the dividend has not yet been dealt with. The result is 144. Looking back at the diagram, pupils can check that 144 is indeed the amount that still has to be divided.

784 ÷ 16			HTU		*16 × table*
			4		1 × = 16
					2 ×
			16 ⟌ 7 8 4		3 ×
			−6 4		4 × = 64
			1 4 4		5 × = 80
	40				6 ×
	640	144			7 ×
16	7̶0̶0̶	8̶4̶			8 ×
		+6̶0̶			9 ×
					10 × = 160

Step 5. Returning to the diagram, pupils turn their attention to the remaining 144. They can refer back to the table skeleton to see where in the table the number 144 falls and can deduce that it is not far short of the 160 entered against 10 × 16. If they are lucky, they will find the answer to 9 × 16 straight away by calculating one step of 16 less than 160, and find that this is the answer they are looking for. If they misjudge it and choose to find the answer to 8 × 16, by doubling the 4 × 16 entry, they will have performed one extra calculation unnecessarily (not a disaster, just extra practice in doubling), and will next have to find 9 × 16. Pupils are now able to mark 9 as the width of the second box.

784 ÷ 16			HTU	*16 × table*
			4	1 × = 16
				2 ×
			16 ⟌ 7 8 4	3 ×
			−6 4	4 × = 64
			1 4 4	5 × = 80
	4 0	9		6 ×
	6 4 0	144		7 ×
16	7̶0̶0̶	8̶4̶		8 ×
		+6̶0̶		9 × = 144
				10 × = 160

Step 6. Turning back to the written notation, pupils are again faced with finding how many 16s are in 144. They now know the answer to be 9, which they record in the units position of the answer. The written workings must now be completed to show that there is no remainder.

(Continued)

(Continued)

784 ÷ 16

```
                          H T U
                            4 9

                      16 | 7 8 4
                          - 6 4
                          -----
                           1 4 4
                          -1 4 4
                          -----
                              0
        40        9
      +------+---------+
      | 6 4 0| 1 4 4   |
   16 | 7̶0̶0̶ | 8̶4̶      |
      |      | +6̶0̶     |
      +------+---------+
```

16 × table	
1 × =	16
2 ×	
3 ×	
4 × =	64
5 × =	80
6 ×	
7 ×	
8 ×	
9 × =	144
10 × =	160

This whole process must be repeated again and again with other problems, until pupils understand the meaning of every step of the written algorithm. If the understanding is there, pupils will not be under pressure to memorise meaningless sequential steps, but can instead work towards the answer from first principles, using either the diagrammatic approach or the standard written algorithms, as they prefer.

Some examples to practise at this stage are: 345 ÷ 15, 1700 ÷ 25, 2415 ÷ 105, 294 ÷ 14, 1416 ÷ 24, 6447 ÷ 21, 5724 ÷ 54, etc. None of these examples result in remainders and all can be sketched on a 2-box rectangle. For example:

5724 ÷ 54

```
                          H T U
                          1 0 6

                      54 | 5 7 2 4
                          - 5 4
                          -----
                            3 2 4
                          - 3 2 4
                          -------
                                0
        100        6
      +-------+---------+
      | 5 4 0 0| 3 2 4  |
   54 | 5̶7̶0̶0̶  | 2̶4̶     |
      |        | +3̶0̶0̶   |
      +-------+---------+
```

54 × table	
1 × =	54
2 ×	
3 ×	
4 ×	
5 × =	270
6 × =	324
7 ×	
8 ×	
9 ×	
10 × =	540

This suggested way of giving pupils some insight into the working of the canonical long division algorithm is not without its problems. In particular, without knowing the size of the answer, pupils cannot reasonably predict how many boxes to make within their rectangle sketch, or how best to split the dividend into sensible chunks. In the example above, for instance, if the pupils had very reasonably chosen to split the 5724 into 5000 + 724 instead of 5400 + 324, the calculation would have been messier and unnecessarily complicated.

An alternative that I have found popular with students of all ages, is to extend the idea of writing out the times table of the divisor, and then proceeding in exactly the same way for

(Continued)

(Continued)

long division as for short division. Taking the same example of 5724 ÷ 54 shown above, pupils can set out the whole times table for 54 as follows: write the skeleton with each step marked and the key facts of 1×, 10×, and 5× entered first, as before; calculate 2×, 4×, and 8× by doubling; calculate 3× by adding the products for 1× and 2 ×; then enter 6× by doubling 3× or by adding one step to 5×; find 9× by calculating one step back from 10× or one step on from 8×. So far, the calculations have not been too taxing and, for pupils who have practised very similar techniques for every times table (see Chapters 5 and 6), should not take long. I often advise pupils to leave a gap for the hardest calculation on the grounds that there is a very good chance that they will not need 7× and if they do they can find it later by adding one step to 6×, or by adding the results for 3× and 4×. Once the times table is in place, pupils can work through the problem as if it were short division. They may need to perform the subtractions separately using column arithmetic, or complementary addition on a number line, if their mental arithmetic is not up to the task.

5724 ÷ 54	54 × table
	1 × = 54
	2 × = 108
	3 × = 162
	4 × = 216
H T U	5 × = 270
1 0 6	6 × = 324
54 ⟌ 5 7³2 4	7 ×
	8 × = 432
	9 × = 486
	10 × = 540

Pupils should be allowed to choose whichever method they feel most comfortable with. Both the diagrammatic boxes method, and the standard written algorithm, are paper-and-pencil methods that show all the pupil's workings, and both can produce the solution in a reasonable amount of time. Providing pupils know how to perform short division confidently, I make a point of not spending too much time and energy on teaching long division. The extra teaching and learning would not contribute to a pupil's further or deeper understanding of the maths, and in the real world everyone would use a calculator.

What to teach next?

See Chapter 9 for how to teach pupils to reason about division in order to simplify numbers wherever possible.

References

1 M. Sharma (various 1980–1993) *Math Notebook*, Center for Teaching & Learning of Mathematics, Framingham, MA, USA. (Professor Sharma's publications and videos are available in the UK from Berkshire Mathematics.)

2 I. Thompson (2003) 'Deconstructing the National Numeracy Strategy's approach to calculation', ch. 2 in *Enhancing Primary Mathematics Teaching*, ed. I. Thompson, Open University Press.

PART IV

Reasoning strategies

CHAPTER NINE

Reasoning strategies

Overview

In a sense, the whole of this book is about reasoning. Certainly, teaching the key strategies in the manner I suggest in the earlier chapters of this book will encourage pupils to think logically and mathematically. Maths is not about finding a correct answer by following a recipe, or about acquiring a page-full of ticks in a school exercise book. It is about making sense of the world, investigating ideas and concepts, making connections, developing and strengthening abstract cognitive skills.

If we acknowledge that pupils should be taught for understanding, we must allow them to limit the facts they have to know by heart to a minimum number of key facts. We must also limit the number of strategies that must be learned to only those key strategies that have the broadest applications. To compensate for the lack of automatic recall, we must encourage pupils to use logic and reasoning to derive what they need to find from what they already know. Pupils who struggle with numeracy at secondary school will not yet have discovered this approach for themselves and therefore need explicit teaching about reasoning strategies.

Summary of reasoning strategies

1. Reasoning about missing numbers.

2. Reasoning from complements to 10 to complements of other numbers.

3. Reasoning from complements to near complements.

4. Reasoning from doubles facts.

5. Reasoning from doubles to near doubles.

6. Reasoning that 9 is almost 10.

7. Reasoning that subtraction can be interpreted as finding the difference.

8. Reasoning from multiplication tables facts.

9. Further reasoning about multiplication.

10. Extending the area model of multiplication to other maths topics.

11. Reasoning about division.

Reasoning

1. Reasoning about missing numbers

Many pupils who can answer a question such as 5 + 7 without difficulty may be defeated by the same problem presented as a missing addend question, i.e. 5 + □ = 12, or the even more difficult formulation of □ + 5 = 12. Pupils who have worked through the activities described in earlier chapters of this book should have acquired a sound understand of the relationship between each of the four operations and its inverse. Pupils who do not make connections between the practical activities and the abstract symbolic notation may need more practice in making and reading and recording equations for themselves. Pupils may also need to be steered away from a purely mechanical process in which a subtraction is turned into an addition. Instead, pupils should practise reasoning by talking to themselves, preferably aloud. Their thinking might go something like this: *The question asks, What must be added to 5 to make 12? Because 12 is the total sum of both numbers, the number I must add must be less than 12. How much less than 12? It must be 5 less than 12.*

Alternatively, pupils could visualise an empty number line and reason as follows: *The question asks, What must be added to 5 to make 12? I can picture 5 on a number line with the number 12 further to the right. The answer to the question lies in the gap between 5 and 12, or the difference between the two numbers, which I can find by bridging up from 5 (through 10) to 12.* This second way of reasoning is connected to the complementary addition approach to subtraction that I believe all pupils should be taught. Note that this second method is by far the best to use if negative numbers are involved.

Similarly, for a missing number subtraction question, pupils should reason their way towards a solution, using language, and preferably aloud. For example, □ − 6 = 15 means *From what number can 6 be taken away to leave 15? The number must be greater than 15. How much greater? It must be 6 more than 15.*

2. Reasoning from complements to 10 to complements of other numbers

Pupils must know by heart the five basic complement facts, because these are key facts. Once known, the same facts can be applied to other units of measurement. For example, the fact that 3 and 7 are complements that combine to make 10 is a basic fact expressed in units of ones. If 3 + 7 = 10, it follows not only that 7 + 3 = 10, but also that 30 + 70 = 100, that 700 + 300 = 1000, that the sum of 3 million and 7 million is 10 million, that 7 metres + 3 metres is 10 metres, that 3 fifteenths plus 7 fifteenths are 10 fifteenths, etc.

Knowledge of complements can be used to derive the sum of any two numbers that add up to a round number (i.e. any multiple of ten). For example, knowing that 3 + 7 = 10 will help towards the solution of 23 + 7, or 583 + 7, or 1027 + 103, etc.

Subtractions from multiples of 10 can be solved by knowing complement facts. For example, knowing that 3 + 7 = 10 allows pupils to solve 10 − 7, or 100 − 30, or 10000 − 3000, or 700 grams less than 1 kilogram, etc. Missing addend problems can be solved in the same way (see also the first strategy in this section, above), for example for finding change from £1, or how far to the finishing line on a track measured in metres or kilometres.

Understanding how one simple fact such as $3 + 7 = 10$ can be extended in so many different directions encourages pupils to generalise. So, if $3 + 7 = 10$, it follows that $3x + 7x = 10x$.

3. Reasoning from complements to near complements

From the five known complement facts, pupils can derive 'near complements'. For example, if $3 + 7 = 10$, it follows that $3 + 8 = 11$ and that $4 + 7 = 11$ and that $3 + 6 = 9$ and that $9 = 2 + 7$.

Coupled with the technique described above of extending complements facts to values other than 10, pupils can derive answers to the questions $0.4 + 0.7$ or four tenths and seven tenths, $123 + 8$, $1070 + 40$, etc.

Near complements facts can be extended, in exactly the same way as described above for complement facts, to subtractions, missing number formulations, fraction work and algebra.

4. Reasoning from doubles facts

From the doubles facts up to $10 + 10$, pupils can derive answers to doubles of greater magnitude. If double 4 is 8, it follows that $40 + 40 = 80$, that $800 - 400 = 400$, that $2 \times 4000 = 8000$, that $2y = 8$ when $y = 4$, etc.

Knowing that $4 + 4 = 8$ means that pupils can know the answer to $24 + 4$, or $694 + 4$, or $2140 + 40$, etc, without counting on and without any real calculation.

Doubles facts help pupils reason about halving. For example, pupils who know that $8 + 8 = 16$ can find half of 16, or $16 \div 2$, or $160 \div 2$, or $160 \div 20$, etc. without any calculation. They can also solve problems such as $216 - 8$ or $4160 - 80$ mentally.

5. Reasoning from doubles to near doubles

If double 7 is 14, it follows that $7 + 8 = 15$, and $6 + 7 = 13$, and $80 + 70 = 150$, and 7 twentieths + 6 twentieths = 13 twentieths, etc.

Reasoning from an understanding of doubles allows pupils to add any two alternating numbers by doubling the middle number. For example, $14 + 16$ is found by doubling 15, and $69 + 71$ can be seen as double 70.

6. Reasoning that 9 is almost 10

The solution to any addition or subtraction problem involving the number 9 can be derived by thinking about 9 as the number that is one less than 10. Some pupils like to think in terms of compensating, others in terms of bridging. For example, to solve the problem $9 + 7$, a compensator would add 10 to 7 and then subtract 1 (but beware of the muddle that often ensues for pupils who use the same technique for both addition and subtraction, and are reduced to making a wild guess at the end about whether to adjust their answer up or down), while a pupil more comfortable with bridging would allow the 9 to take one from the 7 in order to transform itself into the round number 10, leaving 6 to be added. Either kind of reasoning allows pupils to answer questions featuring the number 9, such as 28 + 9, 369 + 5, 24 − 9, 54 − 29, 460 + 90, 821 − 90, etc.

Pupils can reason in the same way about additions or subtractions of the numbers 90, 99, 900, 990, 999, 9000, 9999, etc.

It is also helpful to reason that 9 is almost 10 when the problem is about multiplication, as already explained in earlier chapters. 9 × N is always one step of N less than 10 × N. Pupils can extend this logic to multiplication by 90, 99 or 999, etc.

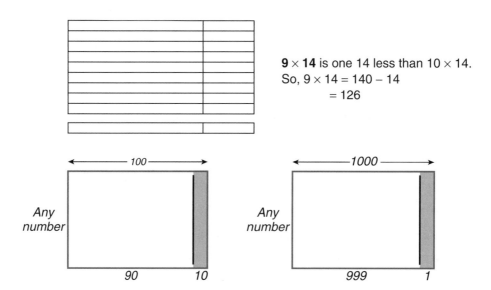

9×14 is one 14 less than 10×14.
So, $9 \times 14 = 140 - 14$
$= 126$

In column subtraction from round numbers, in other words subtractions that would otherwise require successive decompositions, it can be helpful to use the fact that 9 is almost 10 to adjust the question and then compensate for the adjustment at the end. For example, 3000 − 48 can be set out vertically in columns as 2999 − 48, which can be worked through without the need for decomposition. The answer needs to be readjusted by adding 1 at the end.

This is a useful reasoning strategy for those pupils who prefer column subtraction to sequential mental methods when dealing with multi-digit numbers. It can also benefit those who learn to look ahead and judge which subtraction problems might be quicker to solve vertically in columns than by horizontal complementary addition on an empty number line. In this example,

3000 – 48, most people would view the number line as a less attractive way of reaching the solution than the adjusted column subtraction shown above.

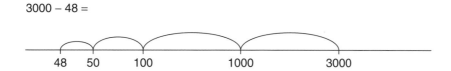

3000 – 48 =

7. Reasoning that subtraction can be interpreted as finding the difference

As described in the first section in this chapter on missing numbers, pupils can reason about subtraction in different ways. The most fruitful way, and the one to be encouraged above all others, is to understand subtraction as a comparison between two numbers on a number line, a question that can be solved by finding the difference between them. This idea is more fully explored in Chapters 2–4 of this book.

Once pupils understand that the answer to a subtraction can be found by calculating the gap between numbers, they can learn to reason that the gap can be shifted up or down the number line at will. The answer to a 'displaced subtraction' remains the same provided both numbers are shifted by the same amount in the same direction. For example, the answer to 43 – 17 is the same as the answer to 53 – 27 or 143 – 117, or 83 – 57, etc. These three new problems have been generated by shifting both numbers of the question, and therefore the fixed gap between the pair of numbers, by one or more 10 or 100, up along the number line. It follows that the answer will also be the same if both numbers are shifted a mere 3 steps up the number line. In this particular example, a shift of 3 would be the optimal choice in order to create the calculation 46 – 20. Clearly, 46 – 20 is much quicker and easier to solve than the original 43 – 17.

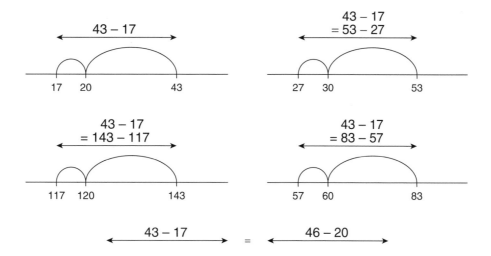

This method for subtraction is sometimes called 'equal additions'. It entails turning the subtrahend (the number that is to be subtracted) into a more congenial number, preferably a multiple of ten, in order to make the calculation easier. For example, a pupil can look at a problem such as 155 – 78 and reason that the answer can be reached by subtracting 80 from 157, or, better still, by subtracting 100 from 177. The aim is to turn a column subtraction problem into a subtraction that can be solved horizontally, either on paper or mentally.

$$\begin{array}{r} 43 \\ -17 \\ \hline \end{array} \xrightarrow[+3]{+3} \begin{array}{r} 46 \\ -20 \\ \hline \end{array}$$

$$\begin{array}{r} 155 \\ -78 \\ \hline \end{array} \xrightarrow[+2]{+2} \begin{array}{r} 157 \\ -80 \\ \hline \end{array} \xrightarrow[+20]{+20} \begin{array}{r} 177 \\ -100 \\ \hline \end{array}$$

8. Reasoning from multiplication tables facts

As has already been thoroughly discussed in earlier chapters of this book, pupils should be taught reasoning strategies about multiplication right from the start. Pupils who cannot reliably remember all the multiplication tables facts should be taught to reason from the key steps of 2, 5 and 10 times the number. Pupils should also be exposed to plenty of practical exercises that allow them to internalise the fact that multiplication is commutative, i.e. that $A \times B = B \times A$.

Therefore, if $6 \times 5 = 30$, it follows that $5 \times 6 = 30$, and that 15×6 will be three times as much as 30, and that 5×12 will be twice 30, and that $6 \times 4 = 30 - 6$ as well as being 6 doubled and doubled again, and that $60 \times 50 = 3000$, and that $300 \div 60 = 5$, etc.

Please see Chapters 5 and 6 for more on reasoning about multiplication tables.

The same logic that is used to derive multiplication facts can be used for larger numbers. The thought process is best supported by sketches of rectangles, as shown here, based on the idea that we have already taught pupils, i.e. that multiplication is best modelled by area. For example, anyone asked to find the answer to 31×21 from the given fact that $21 \times 30 = 630$, may well become confused about whether it is 21, 30 or 31 that must be added to 630. A diagrammatic sketch, like the one shown here, can help to clarify one's thinking and show that what must be added is one 'slice' of 21.

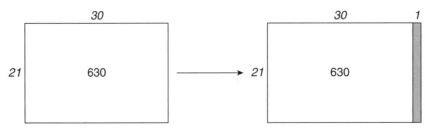

If 21 × 30 = 630, it follows that 31 × 21 = 630 + 21.

9. Further reasoning about multiplication

Refer back to Activity 8 in Chapter 6 to see how pupils can investigate factors and prime numbers by manipulating small cubes in rectangular arrays. The illustration below shows the different shapes and sizes of rectangular arrays that can be created from 16 cubes. Compare this with the single arrangement possible with 17 cubes. Pupils learn to reason from their experience of manipulating concrete materials that numbers forming rectangles can be multiples of other numbers and have factors other than 1 and themselves, whereas those forming only straight lines are prime numbers.

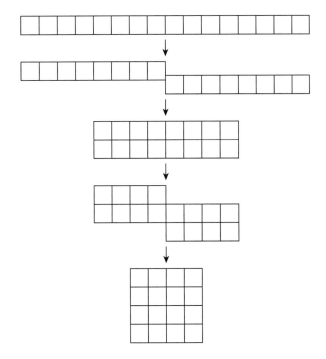

The illustration also shows how concrete materials can be used to show that 16 × 1 is equal to 8 × 2 which is equal to 4 × 4. Having pupils sketch rectangular areas to record the concrete exercise illustrated above will help draw their attention to the fact that if one dimension of the rectangle is halved, the other is doubled. Similar exercises can be carried out with quantities that cannot be easily doubled and halved, for example transforming a 15 × 3 rectangle into one that measures 5 × 9, or vice versa. The general idea is about conservation of area, i.e. that the area remains constant if the measurement of one of the sides is divided by a certain number while the second measurement is multiplied by the same number.

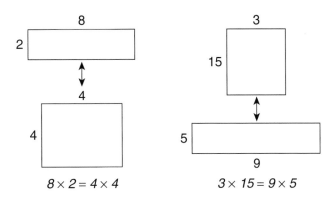

$8 \times 2 = 4 \times 4$ $3 \times 15 = 9 \times 5$

The implications are that some problems that are presented as long multiplication questions can be quickly turned into short multiplication questions. The benefit can be either fewer steps to calculate, or, more often, easier numbers to work with. Pupils with a secure understanding of the area model of multiplication can, by logic and reasoning, turn awkward multiplication questions into accessible ones.

For example, a problem such as 24 × 17 can be sketched on a rectangle and transformed into an equivalent rectangle with 24 ÷ 4 on one side and 17 × 4 on the other, i.e. 6 × 68. This, being a short multiplication question, is already a simpler problem to compute than the original long multiplication of 24 × 17. But it can be made even easier by being turned into either 3 × 136 or 2 × 204.

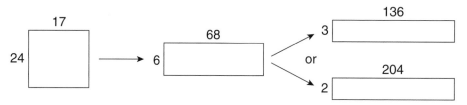

Reshaping a rectangle shows how long multiplication can be turned into short multiplication.

Pupils can reason about how this calculation method works by performing the necessary multiplication in stages. Pupils are able to record the interim answers on a familiar sketch that carries meaning, by virtue of the previous work they have undertaken on the area model of multiplication. Pupils are left with a relatively simple calculation to make to reach the answer. This method can help pupils sidestep multiplication by numbers above 5, which they may find difficult to remember, by substituting several steps involving multiplication by smaller numbers.

10. Extending the area model of multiplication to other maths topics

Once pupils understand that multiplication can be modelled by area and that their thinking about multiplication can be supported by sketches of rectangles, they can reason that the same strategy can be applied wherever multiplication is involved, for example during work on fractions, decimals and algebra.

For example, two fractional numbers multiplied together, or two numbers with decimal places, can be partitioned into convenient chunks, modelled onto a rectangle partitioned into boxes, and calculated in separate steps which are combined to find the answer, as shown here.

Example: $1\frac{1}{2} \times 5\frac{1}{2}$

	5	$\frac{1}{2}$
1	5	$\frac{1}{2}$
$\frac{1}{2}$	$2\frac{1}{2}$	$\frac{1}{4}$

$$1\frac{1}{2} \times 5\frac{1}{2} = 5 + 2\frac{1}{2} + \frac{1}{2} + \frac{1}{4}$$
$$= 8\frac{1}{4}$$

Example: 41.5 × 3.2

	40	1	.5
3	120	3	1.5
.2	8		.3

$$41.5 \times 3.2 = 120 + 8 + 3 + 1.5 + 0.3$$
$$= 132.8$$

Division by a fraction can be treated in just the same way as any other division, by thinking about how to build a rectangle of a given value such that the side of the rectangle measures the same as the given divisor. Working in this way allows pupils to understand what is behind the commonly taught 'trick' of turning the fraction upside down and multiplying it when division is called for. For example, trying to create a rectangle worth 5 such that only $\frac{1}{3}$ is up the side, means that the width of the rectangle has to measure 3 for each one of the 5 units, since there are three thirds in one unit. Therefore, $5 \div \frac{1}{3} = 5 \times 3$.

Example: $5 \div \frac{1}{3}$ Example: $9 \div \frac{1}{2}$

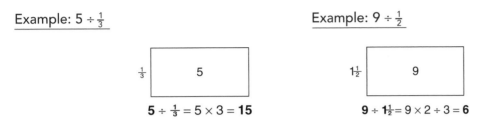

$5 \div \frac{1}{3} = 5 \times 3 = \mathbf{15}$ $9 \div 1\frac{1}{2} = 9 \times 2 \div 3 = \mathbf{6}$

Similarly, expanding algebraic brackets can be treated as any other multiplication and modelled on rectangular sketches. Each multiplier is partitioned in whatever way makes it easier to deal with, so that the partial products can be calculated with confidence before being combined to create the final product. This way of working does not depend on pupils remembering how to perform tricks, like a certain pattern of arrows or a mnemonic such as 'foil'; instead it connects algebraic multiplication to numerical multiplication in a logical way, and demonstrates to pupils visually that when expanding brackets with two terms in each bracket, there will be four partial products to combine. A very common error, in which pupils find only three of the terms, is thereby avoided.

Example: (x + 1) (x + 3)   Example: (2x + 5) (x − 1)

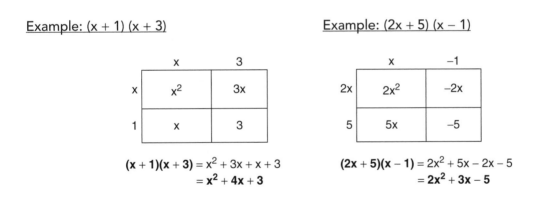

$(x + 1)(x + 3) = x^2 + 3x + x + 3$ $(2x + 5)(x − 1) = 2x^2 + 5x − 2x − 5$
$\qquad\qquad\qquad = x^2 + 4x + 3$ $= 2x^2 + 3x − 5$

11. Reasoning about division

Working with rectangles to model multiplication and division may prompt pupils to notice that division facts have a strong proportional element. If not, this is something we should point out to pupils explicitly. For example, $6 \div 3$ yields the same answer as $12 \div 6$ or $10 \div 5$ or $28 \div 14$, because of the relationship between the divisor and the dividend. The divisor and the dividend are the numbers recorded on the two different sides of a sketched multiplication/division rectangle.

This kind of understanding allows division problems to be treated as if they were fractions that need simplifying. Pupils will first need to be taught that the line between the numerator and denominator of a fraction is a division line. By writing division problems as fractions and cancelling down to the lowest terms, pupils can transform a long division into a problem that can be solved in easy stages using nothing more than basic knowledge of the simpler times tables. Sometimes the whole division problem can be solved in this way; at other times, the cancelling down process results in a fraction with much smaller numbers that can be solved by short division.

For example, problems such as $128 \div 16$ or $384 \div 24$ can be solved much more easily by cancelling down, that is by dividing both numbers in stages, than by long division. A great advantage of this method is that pupils can cancel down in many stages if their knowledge of times tables is insecure, or in a minimum number of stages if their tables knowledge is good. Both the examples $128 \div 16$ and $384 \div 24$, once they are rewritten as fractions, can be cancelled down by dividing by 8 as a first step, or by 4 and then by 2, or by 2 three times.

$$128 \div 16 \longrightarrow \frac{128}{16} \xrightarrow[\div 8]{\div 8} \frac{16}{2} \xrightarrow[\div 2]{\div 2} 8$$

$$\text{or} \quad \frac{128}{16} \xrightarrow[\div 2]{\div 2} \frac{64}{8} \xrightarrow[\div 2]{\div 2} \frac{32}{4} \xrightarrow[\div 4]{\div 4} 8$$

$$384 \div 24 \longrightarrow \frac{384}{24} \xrightarrow[\div 8]{\div 8} \frac{48}{3} \xrightarrow[\div 3]{\div 3} 16$$

$$\text{or} \quad \frac{384}{24} \xrightarrow[\div 2]{\div 2} \frac{192}{12} \xrightarrow[\div 2]{\div 2} \frac{96}{6} \xrightarrow[\div 2]{\div 2} \frac{48}{3} \xrightarrow[\div 3]{\div 3} 16$$

Cancelling down fractions is much easier than performing long division.

A further benefit of this method is when there is a fractional solution. For example, $36 \div 48$ would involve a relatively difficult procedure and a solution with a decimal place if it were treated as a long division problem. The question $36 \div 48$ is much more easily solved by being rewritten as 36 over 48 and cancelled down to its lowest terms. A confident pupil might find the correct result by dividing both top and bottom numbers by 12; an insecure pupil might focus on the fact that both are even numbers, and cancel down by a succession of divisions by 2, i.e. successive halving processes. Either approach would be preferable to having to solve a long division problem in which the divisor is 48.

$$36 \div 48 \longrightarrow \frac{36}{48} \xrightarrow[\div 2]{\div 2} \frac{18}{24} \xrightarrow[\div 2]{\div 2} \frac{9}{12} \xrightarrow[\div 3]{\div 3} \frac{3}{4}$$

Even when the division problem cannot be fully solved by cancelling down, the technique can result in transforming the original question into a much easier problem. For example, $207 \div 63$ is much more troublesome to calculate than $23 \div 7$.

$$207 \div 63 \longrightarrow \frac{207}{63} \xrightarrow[\div 3]{\div 3} \frac{69}{21} \xrightarrow[\div 3]{\div 3} \frac{23}{7}$$

Using the cancelling down technique, the long division example worked through in very great detail at the end of the previous chapter, $784 \div 16$, can be reduced to the much more accessible short division problem of $196 \div 4$ in one or two easy steps (divide both numbers by 4, or by 2 twice). Similarly, an even more daunting example from the end of Chapter 8, $5724 \div 54$ can be easily simplified to $212 \div 2$ (divide both numbers by 3 three times), leaving a simple halving problem to do in order to reach the solution.

To take full advantage of this technique, pupils should be taught the divisibility indicators:

▶ All even numbers are divisible by 2.

▶ Numbers ending in 0 are divisible by 10.

▶ Numbers ending in 5 or 0 are divisible by 5.

▶ If the number formed by the last two digits is divisible by 4, then the whole number is divisible by 4.

▶ If the sum of the digits is divisible by 3, then the number is divisible by 3.

▶ If the sum of the digits is divisible by 9, then the number is divisible by 9.

As an alternative to cancelling down, it is often worth considering the possibility of multiplying up. A problem such as $180 \div 15$ is a good candidate for being treated as a fraction rather than as a long division. Cancelling down, by dividing both top and bottom of the fraction by 5 and then by 3, produces the correct answer with relative ease. However, multiplying both top and bottom of the fraction by 2 as a first step is even quicker because multiplication is easier than division.

$$180 \div 15 \longrightarrow \frac{180}{15} \xrightarrow[\div 5]{\div 5} \frac{36}{3} \xrightarrow[\div 3]{\div 3} 12$$

$$\text{or} \quad \frac{180}{15} \xrightarrow[\times 2]{\times 2} \frac{360}{30} \xrightarrow[\div 10]{\div 10} \frac{36}{3} \xrightarrow[\div 3]{\div 3} 12$$

The cancelling down technique can only work for divisions without remainders, but is nevertheless a very useful division strategy that relies on logical thought and reasoning processes.

Glossary

Addend   A quantity or number that is added to another. For example, in the problem 6 + 8, both 6 and 8 are addends. In the problem 6 + □ = 10, the missing addend is 4.

Algorithm   A prescribed set of rules or sequence of procedures that leads to a solution to a numerical problem.

Bridging   A technique based on a linear understanding of the number system, by which addition or subtraction is performed not in single steps but in convenient chunks or jumps. A numerically significant number, such as 10 or multiples of 10 (in a decimal number system), functions as a stepping stone between the jumps.

Cancelling down   Reducing a fraction to a simplified but equivalent quantity by dividing both the top and bottom numbers by the same amount. Fractions can be cancelled down only if the numerator and the denominator have a common factor.

Chunking   Grouping together a small number of items.

Commutative property   The characteristic of an operation in which the numbers can be taken in any order without altering the result. For example, addition is commutative because $3 + 5 = 5 + 3$ and multiplication is commutative because $3 \times 5 = 5 \times 3$.

Complement   The quantity or number that completes another. In this book, it is usually 10 or multiples of 10 that are considered to be the 'perfect' number that needs to be made whole. For example, 2 is the complement of 8, so 8 and 2 are complements to (or complements of) 10, and 22 and 78 are complements to/of 100.

Complementary addition   A method for subtraction that is based on the idea of complements, or completing a number, by working forwards to find the difference between two numbers or quantities. For example, the solution to 25 − 18 is found by counting, or preferably bridging, up from 18 to 25.

Component   A constituent part of a larger whole. In this book, the word component is often used as a way of distinguishing between a chunking approach and a ones-based approach. For example, the components of 6 are 3 and 3 (or 2 + 4, or 5 + 1, or 1 + 2 + 3). By contrast, a ones-based approach regards 6 as if it were only ever built from 1 + 1 + 1 + 1 + 1 + 1.

Decomposition   Breaking down a number into constituent elements, for example into place value components. Decomposition would be needed in a problem such as 21−15 if it were solved by column subtraction: subtracting each column value separately cannot be achieved until one of the tens is decomposed into units.

Divisor/dividend   The dividend is the quantity or number that is divided by the divisor. For example, in the problem $143 \div 11$, 143 is the dividend and 11 is the divisor. (The solution, 13, is the quotient.)

Empty number line   See 'number line'.

Factor   A number or a quantity that divides exactly into another. For example, 5 is a factor of 15. (The other factors of 15 are 1, 3 and 15.)

Mental arithmetic   Numerical calculation by mental processes, as opposed to solutions resulting from written procedures. Seeing numbers written down, and jotting down interim solutions, should be regarded as acceptable memory aids to support mental calculation.

Minuend   A quantity or a number from which another is subtracted. For example, in the problem $25 - 7$, 25 is the minuend. (7 is the subtrahend.)

Multiplier/multiplicand   The multiplier and the multiplicand are quantities or numbers that are multiplied together. Since multiplication is commutative, it does not really matter which is which. For the record, in the problem 4×8, 4 is the multiplier and 8 is the multiplicand.

Nuggets   Small polished hemispherical pieces, often made of coloured plastic or iridescent glass and usually sold as vase fillers or table decorations. They make an attractive alternative to counters.

Number line   A much more abstract linear model of the number system than a number track (see below). A number line is a line on which numbers are represented as points on the line. Using a number line for counting on means counting the intervals between numbers, not the numbers themselves. Number lines can be labelled in a great variety of ways and can cater for fractional numbers as easily as for whole numbers. Any labelled quantities are not necessarily consecutive numbers.

The most versatile number line for supporting mental arithmetic is the empty number line, on which no number is labelled in advance of a calculation being carried out. Informal jottings are made on the line to support mental calculation, with no requirement to keep to scale and with the freedom to give number labels to as many or as few points on the line as desired. Addition or subtraction is represented as movement along the line and sketched above the line as arcs to illustrate component 'jumps'.

Number track   A model of the number system that allocates a defined space, or area, to each number. Whether or not the number track is fully or partially labelled, each of the equal spaces is allocated to a single number only, with abutting cells representing adjacent numbers. Number tracks are only suitable for consecutive whole numbers.

Partition   To split into separate segments or component parts. When a number is partitioned it is split into smaller quantities, usually to make a calculation easier to perform. The act of partitioning a number creates two or more components. For example, 12 can be partitioned into 6 and 6, or into 10 and 2, or into 3 fours, etc.

Place value   The system in which a digit's value is determined by its place or position within a number. For example, in the number 555 each of the fives carries a different value.

Round number   A colloquial expression often used by younger children for any number ending in zero. The correct mathematical term is 'multiple of 10'.

Subitise   The ability to perceive a quantity without having to count it. Most people can subitise amounts of up to 4 easily, with a normal human limit of up to 7.

Subtrahend   A quantity or a number that is subtracted from another. For example, in the problem 25 − 7, 7 is the subtrahend. (25 is the minuend.)

Suggestions for further reading

Julia Anghileri (2000, new edn 2006) *Teaching Number Sense*, Continuum.

Ronit Bird (2007) *The Dyscalculia Toolkit*, Sage.

Brian Butterworth (1999) *The Mathematical Brain*, Macmillan.

Brian Butterworth and Dorian Yeo (2004) *Dyscalculia Guidance*, nferNelson, David Fulton.

Stephen Chinn (2004) *The Trouble With Maths*, Routledge Falmer.

Steve Chinn and Richard Ashcroft (1998) *Mathematics for Dyslexics: A Teaching Handbook*, Whurr.

Stanislas Dehaene (1997) *The Number Sense*, Penguin.

Eva Gravberg (1997) *Elementary Mathematics and Language Difficulties*, Whurr.

Derek Haylock (1995, new edn 2006) *Mathematics Explained for Primary Teachers*, Paul Chapman Publishing.

Tim Miles and Elaine Miles (eds) (1992, new edn 2004) *Dyslexia and Mathematics*, Routledge.

Mahesh Sharma (variously 1980–93) *Math Notebook*, Center for Teaching & Learning of Mathematics, Framingham, MA. (Professor Sharma's publications and videos are available in the UK from Berkshire Mathematics.)

Ian Thompson (ed.) (1997) *Teaching and Learning Early Number*, Open University Press.

Ian Thompson (ed.) (2003) *Enhancing Primary Mathematics Teaching*, Open University Press.

Dorian Yeo (2003) *Dyslexia, Dyspraxia & Mathematics*, Whurr.

Index